Nutrients
and Immune
Function

Nutrients

and Immune

Function

HENG FONG SEOW

Universiti Putra Malaysia, Malaysia

World Scientific

NEW JERSEY · LONDON · SINGAPORE · BEIJING · SHANGHAI · HONG KONG · TAIPEI · CHENNAI · TOKYO

Published by

World Scientific Publishing Co. Pte. Ltd.

5 Toh Tuck Link, Singapore 596224

USA office: 27 Warren Street, Suite 401-402, Hackensack, NJ 07601

UK office: 57 Shelton Street, Covent Garden, London WC2H 9HE

Library of Congress Cataloging-in-Publication Data

Names: Seow, Heng Fong, author.

Title: Nutrients and immune function / Heng Fong Seow, Universiti Putra Malaysia, Malaysia.

Description: 1st edition. | New Jersey : World Scientific, [2021] |
    Includes bibliographical references and index.

Identifiers: LCCN 2020032157 | ISBN 9789811225888 (hardcover) |
    ISBN 9789811227707 (softcover) | ISBN 9789811225895 (ebook) |
    ISBN 9789811225901 (ebook other)

Subjects: LCSH: Immunity--Nutritional aspects.

Classification: LCC QR182.2.N86 S46 2021 | DDC 616.07/9--dc23

LC record available at https://lccn.loc.gov/2020032157

British Library Cataloguing-in-Publication Data

A catalogue record for this book is available from the British Library.

For any available supplementary material, please visit
https://www.worldscientific.com/worldscibooks/10.1142/11984#t=suppl

Desk Editor: Shaun Tan Yi Jie

Typeset by Stallion Press
Email: enquiries@stallionpress.com

Preface

Nutrition plays a fundamental role in the maintenance of good health. The link between nutrition and immunity is evident from the effects of malnutrition which results in secondary immunodeficiency. It has also been well known for decades that vitamin and mineral deficiencies can lead to an increase in susceptibility to infectious diseases. In recent years, advances in clinical and experimental studies have provided the scientific basis of the effects of nutrients on immune cells. The effects of micronutrients, probiotics and prebiotics on the immune system are mostly based on data from experimental studies in animal models unless otherwise specified. This field of research has progressed to those with increased complexity involving skilled knowledge in immunology, molecular and cell biology.

This book is aimed at nutrition or dietetics students who are learning immunology for the first time. A basic introduction to the components and functions of the immune system is described in Chapter 1. Micronutrients such as vitamins and minerals play a critical role in maintaining health and well-being. The impact of these micronutrients on immune function is described in Chapters 2 and 3. Then, the link between inflammation and diseases, and how polyunsaturated fatty acids can play a role in resolving inflammation are explained in Chapter 4. Food allergies are a common occurrence and how the immune system deals with food allergens is described in Chapter 5. Specific immune responses at the gastrointestinal tract and effect of probiotic consumption on immunity are explained in Chapter 6. In recent years, there has been a lot of progress in understanding the cause of autoimmune diseases. The role of nutrition and diet in autoimmune diseases is described in Chapter 7. The role of nutrition and diet in cancer risk reduction and immunonutritional support for cancer patients are described in Chapter 8. The effects of exercise on the immune system are described in Chapter 9. Immunonutritional support for the elderly and critically ill patients is described in Chapter 10.

This book will provide the fundamental knowledge essential for nutrition and dietetics students learning immunology. It is aimed at building a good foundation for students to understand the immunological basis of how micronutrients can affect the function of cells of the immune system to maintain a healthy body in the normal state. Readers will also obtain a better understanding of the scientific basis of nutrition in disease risk reduction and as immunonutritional support in elite athletes, the elderly and critically ill patients.

Contents

Chapter 1

Elements of the immune system

Learning objectives

After studying this chapter, you should be able to:

1. Explain the organisation of the immune system, innate and adaptive immunity
2. State the basic functions of the cells of the immune system and their origin
3. Explain the role of primary and secondary lymphoid organs
4. Explain the process of antigen recognition, processing and presentation to the T cell receptor and B cell receptor
5. State the cytokines involved in the differentiation of various T helper cell subsets
6. Explain the functions of these T helper subsets
7. Explain the gastrointestinal immune system

The immune system is the defence system of the body consisting of cells, tissues and molecules that defend the body against invading micro-organisms such as bacteria, viruses, fungi and parasites and is tolerant to non-threatening organisms, food components and self-antigens. This complex network of cells, tissues and organs work together to maintain homeostasis and eliminate foreign antigens. However, immune responses to allergens can give rise to allergic conditions and immune reactions to self-antigens result in autoimmune diseases which are unfavourable to the host.

There are many examples of nutrient deficiencies that can result in immunodeficiency, leading to increased susceptibility to infectious diseases. The effects of specific nutrient deficiencies on immune function have been investigated mainly by use of animal models. Studies on nutritional immunosuppression in humans are complicated by factors such as environmental circumstances (e.g. poverty, poor sanitation) and gene polymorphisms. In order to understand the effects of nutrients and nutrient deficiencies on the immune function, one needs to learn about the organisation and basic elements of the immune system, which is the focus of this chapter.

1.1 Organisation of the immune system: Lymphatic system

The lymphatic system, which is closely related to the blood circulatory system, contains lymphatic vessels, lymphatic capillaries, lymph nodes and the spleen. It carries lymph which is a clear, watery fluid that contains substances including proteins, salts, and white blood cells throughout the body and towards the heart for circulation in the blood. It is an essential part of the immune system as the extensive drainage network for defence against foreign antigens and removal of toxins, wastes, excess fluids, and infectious agents from all tissues in the body. Foreign antigens from an infected site reach the lymph node via the lymphatic system (Figure 1.1).

Cells of the immune system such as T lymphocytes (T cells) and B lymphocytes (B cells) originate in the thymus and bone marrow, respectively, and circulate to peripheral tissues through the blood and lymph. T cells develop in the thymus whilst B cells develop in the bone marrow, and they interact with antigens in the lymph nodes and spleen.

1.2 Primary lymphoid organs

The thymus is the primary lymphoid organ for T cell development. The bone marrow is the primary lymphoid organ for B cell development. Once matured, they exit the primary lymphoid organs and circulate through the body via lymph and blood.

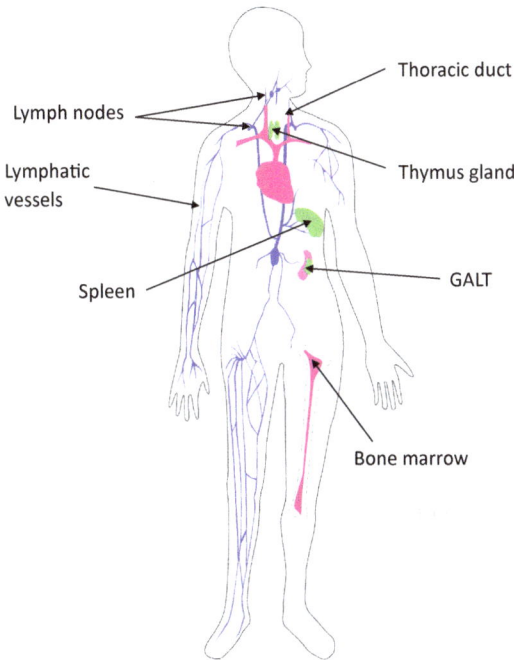

Figure 1.1   Primary and secondary lymphoid organs and the lymphatic system.

Key points on lymphoid organs:

- Primary lymphoid organs: thymus and bone marrow
  — provide environment for lymphocyte maturation
- Secondary lymphoid organs: spleen, lymph nodes, and mucosal-associated lymphoid tissue
  — provide microenvironment for interaction between lymphoid cells and trapped microorganism

---

1.3 Secondary lymphoid organ: Lymph nodes

The lymph node and spleen are referred to as secondary lymphoid organs. Foreign antigens from tissues enter the lymph nodes while those in the blood stream enter the spleen. Where are lymph nodes found? Lymph nodes are distributed throughout the body in groups along the larger lymphatic vessels. The major clusters of lymph nodes are found in six areas, namely, cervical, arm pits (axillary), inguinal, pelvic, abdominal (mesenteric) and thoracic (mediastinal) regions (Figure 1.2). As shown in Figure 1.3, antigens of invading pathogens are captured by dendritic cells

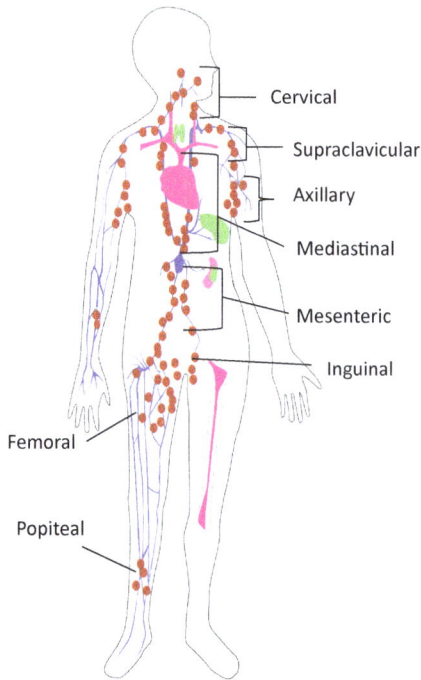

Figure 1.2   Location of lymph nodes.

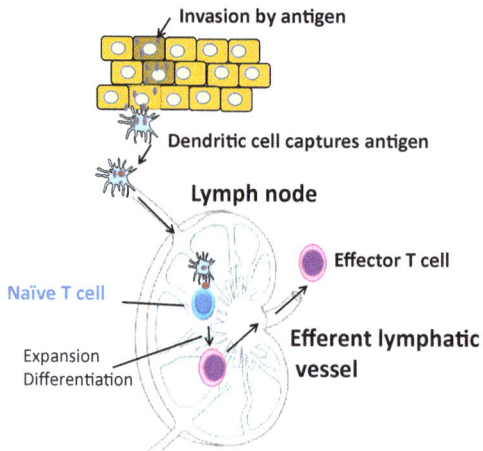

Figure 1.3   Trafficking of dendritic cells to the draining lymph node after antigen capture.

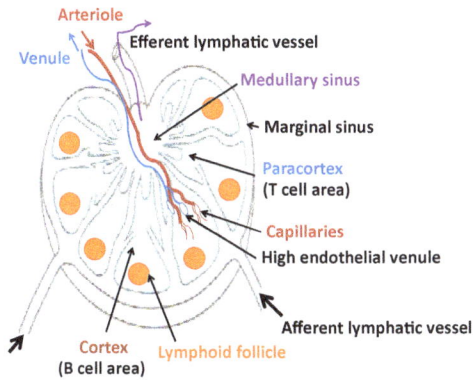

Figure 1.4   Schematic representation showing anatomy of the lymph node.

and transported from tissues via lymphatic vessels to the lymph nodes, where they encounter immune cells such as T and B cells.

Lymph nodes are specialised to trap antigen-bearing dendritic cells, present processed antigens to T cells and allow initiation of adaptive immune responses. A lymph node can be divided into three zones, namely, the cortex, paracortex and medulla (Figure 1.4).

B cells and follicles are found in the cortex. The follicles found in the cortex can contain germinal centres, which is a site for antibody production, somatic hypermutation and antibody isotype switching. The paracortex contain T cells in association with dendritic cells. Macrophages and plasma cells are located in the medulla. The steps involved in transporting antigens to the draining lymph node is as shown in Figure 1.3. Antigens captured by dendritic cells drain into the lymph node via the afferent lymphatic vessel. In the lymph node, antigen is processed and presented to the T cell receptor (TCR) in association with the major histocompatibility complex (MHC) protein. Effector lymphocytes, plasma cells, memory cells and antibodies exit the lymph node via the efferent lymphatics.

1.4 Secondary lymphoid tissue: Spleen

The spleen is a soft encapsulated organ which lies beneath the rib cage and above the stomach in the left upper quadrant of the abdomen (Figure 1.5A).

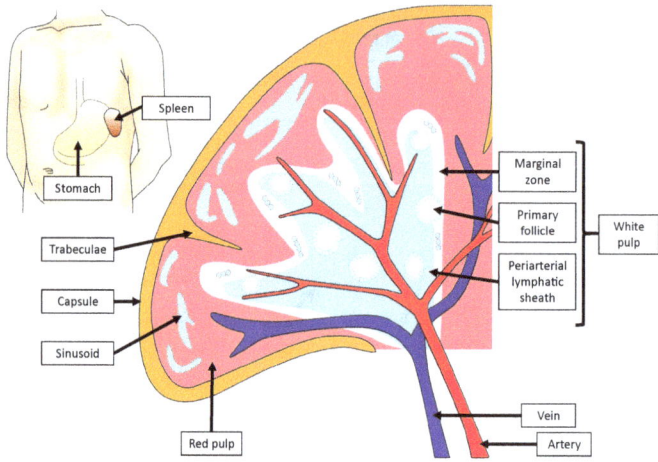

Figure 1.5   (A) Location of the spleen in the body. (B) Schematic representation showing anatomy of the spleen.

It has a rich blood supply from the splenic artery and has the red and white pulp regions (Figure 1.5B). The red pulp is the site for clearance of immune complex and recycling of the red blood cells. The white pulp is the site for antibody production, antigen presentation and lymphocyte activation similar to the functions of the lymph node. Antigens can enter from the bloodstream into the spleen. The spleen has the ability to filter bacteria from the blood circulation.

1.5 Cytokines as communicators of the immune system

Cytokines are small secreted peptides that play a key role in intracellular communication between cells of the immune system. An inducing stimulus switches on cytokine gene expression which is released into surrounding tissues. Cytokines can turn on or off immune responses and mediate inflammation.

They are present at low concentrations. Their functions include stimulation of cell growth and proliferation such as interleukin (IL)-2, inhibition of T cell differentiation for example IL-10, and promotion of inflammation by tumour necrosis factor (TNF)–α and IL-1β. Soluble forms of cytokine receptors such as IL-1 receptor antagonist and soluble TNF receptors inhibit the activity of IL-1β and TNF, respectively.

Cytokines act both locally and systemically (local action — affect cells in very close proximity; systemic — affect organs far from site of infection). Cytokines are pleiotropic, that is, they show many biological activities. Chemokines is a term used to refer to cytokines that have chemoattractant ability, that is, they attract cells to the site of infection. For example, IL-8 serves as a chemoattractant to induce neutrophils to move to the site of infection/inflammation.

Cytokines were first described in the 1970s. They were named according to their biological activity. There was confusion when it was found that the same protein with different biological activities was referred to by different names. Thus, a nomenclature using the term 'interleukin followed by a number' was adopted to avoid confusion.

However, TNFs, interferons (IFNs) and the colony stimulating factors (CSFs) retained their original names. TNF was named according to its ability to cause necrosis of tumours in animal studies. It is produced in large quantities in response to bacterial lipopolysaccharides.

Type I IFN-α and -β are produced in response to viral infection and they act to limit the spread and growth of viruses. IFN-γ is an important activator of phagocytes, and is also important in viral and tumour immunity. CSFs such as granulocyte-macrophage CSFs (GM-CSFs) are growth factors which induce differentiation of cells such as haemopoietic progenitor cells in the bone marrow.

1.6 Mechanism of action of cytokines

Cytokines exert their action on cells by binding to a cell surface receptor, triggering the signal transduction cascade and activation of gene expression. Only cells with receptors specific to the cytokine will respond (Figure 1.6). Examples are as follows: (1) IL-1 or TNF have receptors for these cytokines on many different cell types and therefore have wide-ranging effects. (2) IL-2 only acts on T and B cells so the IL-2 receptor is expressed on these cells only. Activated T cells produce IL-2 which is well-known as a T cell growth factor. When IL-2 binds to IL-2 receptor on T cells, the T cells proliferate and expand.

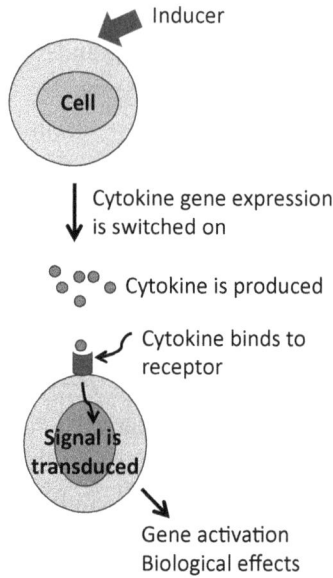

Figure 1.6   Mechanism of cytokine action. Cytokine is produced by a cell in response to a stimulus. The cytokine binds to its receptor, triggering the signal transduction cascade and activation of genes leading to biological effects such as cell proliferation.

1.7 Role of adhesion molecules in immune responses

Cell-cell contact occurs at many stages of immune responses. Expression of adhesion molecules controls cell-cell contact, which are important in many cellular processes such as cell activation, cell migration, homing and co-stimulation.

Cell-to-cell adhesion is critical in the movement of cells from the blood into the site of inflammation or infection. Cells such as neutrophils need to make contact with the vascular endothelium. Rolling and adhesion of neutrophils requires contact between Sialylated Lewis X antigen on neutrophils with E-Selectin on the endothelium. Firm adhesion involves the transient interaction between adhesion molecules of the vascular endothelium such as intercellular adhesion molecule I (ICAM-1) and leucocyte function antigen-1 (LFA-1) on neutrophils, followed by transmigration of the neutrophils across the endothelium into the site of infection

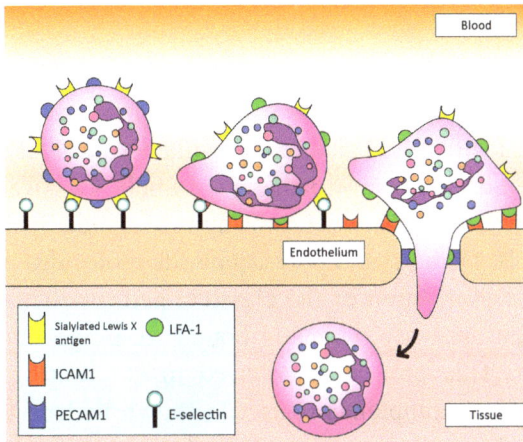

Figure 1.7   Interaction between adhesion molecules of neutrophils and endothelium.
Sialylated Lewis X antigen on neutrophil interacts with E-Selectin on the endothelium.
Interaction between intercellular adhesion molecule I (ICAM-1) on endothelium and leucocyte function antigen-1 (LFA-1) on neutrophils facilitates firm adhesion of neutrophils,
leading to transmigration of neutrophils from the blood into the site of infection.

(Figure 1.7). Expression of these adhesion molecules are induced by
cytokines such as TNF and chemokine CXCL8.

Adhesion molecules and chemokines also play a role in the homing of
naïve T cells to secondary lymphoid tissues as well as interaction between
antigen presenting cells (APCs) and T cells. Homing is the process by
which naïve T cells from the blood enter the T cell area of the lymph node.
Naïve T cells expressing CCR7 (whose ligands are CCL21 and CCL19)
home to the T cell area of the lymph node in response to CCL21 and
CCL19 secreted by stromal and dendritic cells there. Interaction between
adhesion molecules allows the initial contact between the T cell and the
vascular endothelium. Next, T cells bind to the dendritic cell via interaction between ICAM-1 on the dendritic cell and LFA-1 on the T cell,
resulting in stabilisation of recognition by the TCR and the specific antigenic peptide complexed to MHC. Gut-associated DC increases the homing of B and T lymphocytes to the small intestine in the presence of the
vitamin A metabolite, retinoic acid, via upregulating the expression of
chemokine receptor 9 and integrin $\alpha4\beta7$.

1.8 Complement

The complement system consists of at least 20 proteins that are present in the blood circulation in inactive forms. The proteins include complement C1 to C9, factors B, D and P and regulatory proteins. Complement can be activated by two major pathways, namely, classical and alternative. Activation of the alternative pathway is triggered by the interaction among factors B, D and P and polysaccharide molecules present on the microorganisms. Activation of the classical pathway is triggered when C1 binds to the antibody which is bound to antigen present on the microorganism. This process is referred to as complement fixation. Components of the complement cascade shown include C1q, C1r and C1s, which are the first component of the classical complement pathway acting as an enzyme complex which cleaves C4 and C2. C4b binds to the cell surface and complexes with C2a, which then acts on C3. Both classical and alternative pathways converge at C3, which is cleaved to C3a and C3b (Figure 1.8). C3b initiates the formation of a membrane attack complex, which causes cell lysis. C-reactive protein (CRP) can also activate complement. CRP is produced by the liver in response to

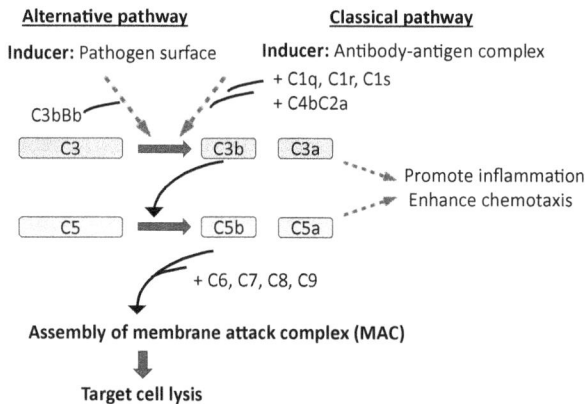

Figure 1.8   Alternative and classical complement pathways. Both alternative and classical pathways converge at C3, which is cleaved to C3a and C3b. A cascade of complement proteins is cleaved, namely, C5, C6, C7 and C8, leading to the assembly of the membrane attack complex consisting of polymeric C9, which inserts into the cell membrane and causes cell lysis.

inflammation, and is a clinical biomarker for acute infection and inflammatory conditions.

1.9 Lines of defence in the immune system

The immune system has three lines of defence. The first line consists of physical barriers such as the skin and mucosal surfaces of the gastrointestinal, urinogenital and respiratory tracts. These are strategically located and they function to prevent entry of pathogens. They immediately attack and destroy the infectious agents that have entered the body.

For example, mucus and cilia found in the nose and throat can trap the pathogens that try to enter the body, and they are pushed out via coughing and sneezing. At the epithelial cell surface, there is a rapid turnover and shedding of epithelial cells. Pathogens that stick to the surface of the epithelial lining will be slough off and thus, entry of pathogen is stopped. Keratin in the skin presents a barrier to most pathogens.

Those that escape the first line will be attacked by cells of the innate immune system such as the phagocytes, proteins such as enzymes in saliva and tears, and complement. There are also chemicals such as lysozyme, defensins and lactoferrin that will kill the invading pathogen. Stomach mucosae secrete acid and digestive enzymes. Skin acidity and chemicals such as fatty acids can kill certain bacteria. Thus, anti-microbial proteins, phagocytes and other cells of the innate immune system contribute to the second line of defence and they function to block the spread of the invading pathogen throughout the body.

The cells and molecules of the innate system are ready to attack within the first few hours of entry. The word innate refers to what you are naturally born with. Thus, cells such as phagocytes and proteins such as enzymes, complement, clotting factors and cytokines are naturally existing and ready to act when the pathogen crosses the first line of defence.

Note that there are commensal organisms that colonise the host mucosal surfaces and they are good organisms as they physically occupy the mucosal surfaces and prevent the entry of harmful pathogens. However, antibiotics that are administered to treat bacterial infections can kill the commensal bacteria. Therefore, proper clinical management of antibiotics is important.

The third line of defence is adaptive immune system. It consists of T and B cells. T cells consist of those that can kill and those that provide help to B cells to produce antibodies. The T and B cells take longer to react as it will take a number of days or weeks for them to be ready. They work in conjunction with the innate immune system. The features of adaptive immunity is antigen specificity, diversity, immunological memory and the ability to distinguish between self and non-self.

1.10 Origin of cells of the immune system

The cells of the immune system originate from stem cells in the bone marrow. These stem cells in the bone marrow differentiate either along lymphoid or myeloid lineage.

Lymphoid progenitor cells give rise to T and B lymphocytes and natural killer cells. The myeloid progenitor cells give rise to erythrocytes, monocytes, neutrophils, mast cells, basophils, eosinophils, and megakaryocytes which give rise to platelets (Figure 1.9).

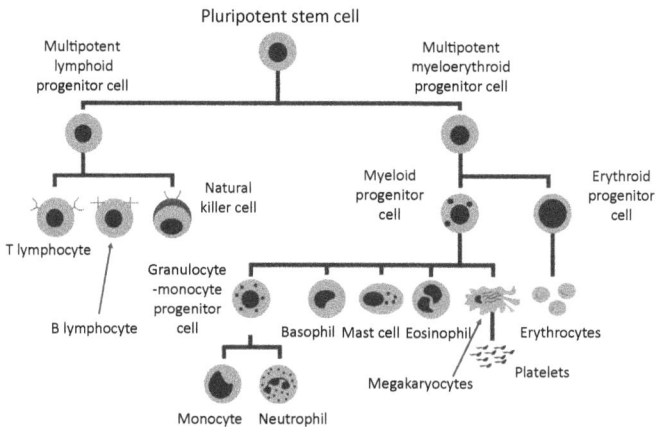

Figure 1.9   Origin of blood cells. Lymphoid progenitor cells in the bone marrow give rise to T and B lymphocytes. Myeloid progenitor cells give rise to erythrocytes, monocytes, neutrophils, mast cells, basophils, eosinophils, and megakaryocytes which give rise to platelets.

The bone marrow is one of the tissues in the body which is most metabolically active. Billions of blood cells are produced daily. Adequate intake of nutrients is important for production of immune system components.

1.11 Cells of the innate immune system

These cells include neutrophils, macrophages, natural killer cells, esoinophils and basophils.

They respond rapidly to all types of threat such as pathogens, injury, stress and destroy the threat by engulfing and destroying or by releasing proteins that can directly kill the target or bring in other immune cells. They communicate directly with the adaptive arm of immunity to initiate a primary immune response.

1.11.1 *Neutrophils*

The most abundant leukocyte in blood is the neutrophil. It has a multi-lobed nucleus and has various granules. The granules consist of enzymes as shown in Figure 1.10A. The surface of the neutrophil harbours receptors for complement, cytokines, adhesion molecules, antibody and Toll-like receptors (TLRs) (Figure 1.10B). Neutrophils engulf the bacterium and a phagosome is formed. Fusion of lysosome with phagosome and digestive enzymes kills the ingested pathogen. After taking up microorganisms, neutrophils undergo respiratory burst (oxidative burst). When neutrophils perform oxidative burst, superoxide anion radicals are produced from oxygen in a reaction linked to the oxidation of glucose (Figure 1.11). The reactive oxygen species produced can be damaging to host tissues and thus antioxidant protective mechanisms are required to avoid damage. Bacteria are killed by reactive oxygen species such as superoxide and enzymes such as lysozyme. The neutrophil dies after killing the pathogen and is removed by macrophages.

The antioxidant protective mechanisms required to avoid tissue damage can be provided by classic antioxidants which include vitamins (vitamins E and C), glutathione, the antioxidant enzymes superoxide dismutase and catalase, and the glutathione recycling enzyme glutathione peroxidase.

Azurophil granules
Myeloperoxidase
Neutral serine proteases
  cathepsin G
  elastase
  proteinase 3
Bacterial/permeability-
  increasing protein
Defensins
Lysozyme

Specific granules
Lactoferrin
Lysozyme
Cytochrome b558
Collagenase
Gelatinase
CD11b/CD18
fMl P-R

Secretory granules
Cytochrome b558
CD11b/CD18
CR1
Alkaline phosphatase
fMLP-R

Gelatinase granules
Lysozyme
Cytochrome
b558Gelatinase
CD11b/CD18
Acetyltransferase

(A)

Antibody or
antibody-opsonised
particles

Fc receptors

C5a, C3bi or
complement-
opsonised particles

Complement receptors

Cytokine receptors
Eg. IL-1R, G-CSFR,
TNFR

Toll-like receptors
Eg. TLR1, 2, 4, 9

Cytokines eg. IL-1, G-
CSF, TNF

Pathogen products eg. LPS

Cell-surface molecules
Eg. intercellular
adhesion molecule-1
(ICAM-1), E-selectin.

Adhesion molecules
Eg. lymphocyte-associated antigen-1(LFA-1),
Sialylated Lewis X antigen.

(B)

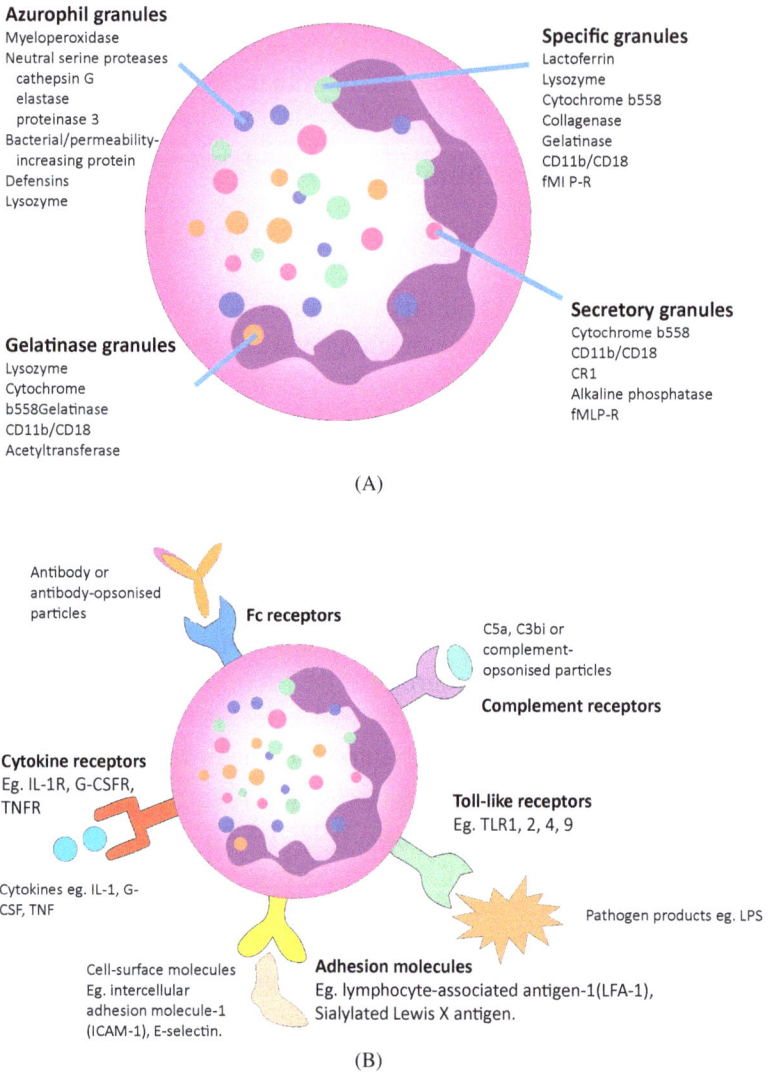

Figure 1.10   (A) Schematic representation showing content of neutrophil granules and (B) cell surface receptors of neutrophils.

These antioxidant enzymes have metal ions at their active site. Cytosolic superoxide dismutase and catalase have copper in their active site; glutathione peroxidase has selenium in the active site. Effects of mineral deficiency are explained in Chapter 3.

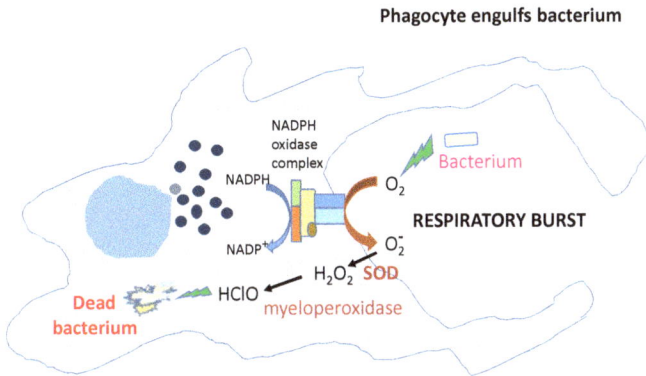

Figure 1.11   Respiratory burst by phagocyte. Engulfment of bacteria triggers respiratory burst resulting in release of reactive oxygen species to kill engulfed bacteria.

1.11.2 *Monocytes*

Monocytes circulate in blood. A monocyte has a U-shaped nucleus and no prominent granule. They migrate into the tissues and differentiate into macrophages which can take up antigens. Resident macrophages in various tissues are known by different names as follows: Kupffer cells in liver, alveolar macrophages in lung, microglial in central nervous system and osteoclasts in bone.

Macrophages have carbohydrate receptors, such as N-formyl methionyl receptor which recognise antigens on pathogens. They also have TLRs which recognise pathogen-associated molecular patterns. Activated TLRs trigger the release of cytokines that promote inflammation (Figure 1.12).

1.11.3 *Dendritic cells*

Dendritic cells are professional APCs which reside in tissues as well as in the blood circulation (Figure 1.13). They engulf pathogens, process them into peptides and present the antigenic peptides to the TCR on the surface of T cells.

1.11.4 *Mast cells and basophils*

Mast cells and basophils reside in many tissues. They have important roles in host defence and inflammation.

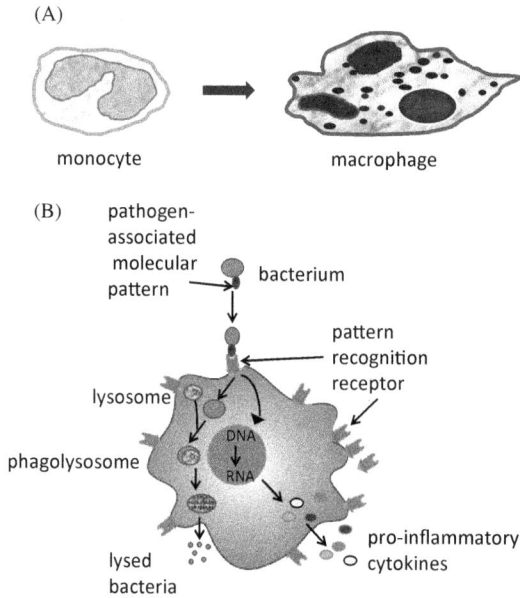

Figure 1.12 (A) Schematic representation of blood monocyte and tissue macrophage. (B) Effect of binding of pathogen-associated molecular patterns to pattern recognition receptor such as TLR.

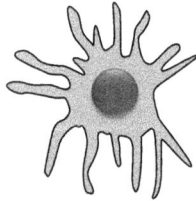

Figure 1.13 Schematic representation of a dendritic cell.

Mast cells are packed full of dense granules containing toxic substances to kill targets and proteins to trigger inflammatory responses (Figure 1.14). These granules include histamine and other inflammatory proteins or chemicals. Once activated, mast cells are triggered to release them. Mast cells also play an important role in asthma.

A basophil has blue-staining granules, and contain chemicals that also mediate inflammation in allergy.

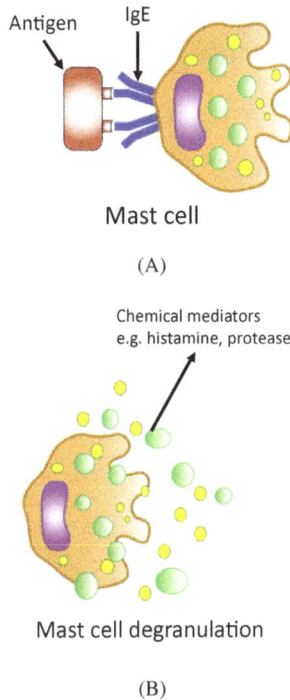

Mast cell

(A)

Mast cell degranulation

(B)

Figure 1.14   Mast cells. (A) Antigen-specific IgE on mast cells bind to antigen. (B) Antigen binding to IgE bound to Fcε receptor triggers mast cell degranulation.

1.11.5 *Eosinophils*

Eosinophils are important in defence against helminth parasites. An eosinophil has a bi-lobed nucleus and prominent red-staining granules. These granules contain cationic proteins and enzymes which kill the parasites (Figure 1.15).

1.11.6 *Natural killer cells*

Natural killer cells are granular lymphocytes that react to eliminate cancerous and virus-infected cells. When triggered to kill their targets, they release perforins and granzymes which lyse the target cells (Figure 1.16).

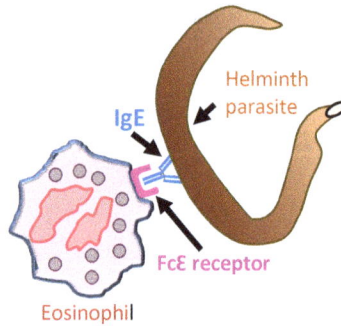

Figure 1.15 Recognition of IgE-bound to helminth parasite by Fcε receptor on eosinophil.

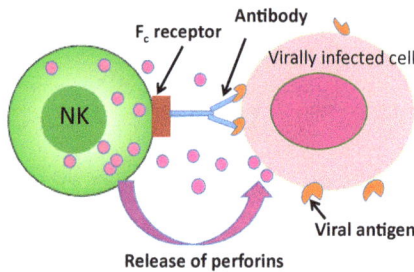

Figure 1.16 Antibody bound to FcΥ receptor on natural killer cell binds to viral antigen on virally infected cell and triggers activation of NK cell.

1.12 Cells of the adaptive immune system

The adaptive immune system has two arms, namely, humoral (antibody-mediated) and cell-mediated immunity (CMI). There are two types of lymphocytes, namely, B lymphocytes (B cells) and T lymphocytes (T cells). B cells play a key role in humoral immunity and T cells constitute CMI. These cells have a round nucleus, no prominent granules and are smaller than a monocyte. Immature T and B cells originate from the bone marrow. B cells become immunocompetent in the bone marrow. Some self-reactive B cells are inactivated or killed. Other B cells undergo receptor editing in which their receptors are rearranged. T cells mature in the thymus. The process of positive selection selects for T cells which become immunocompetent and self-tolerant. Negative selection eliminates T cells that are strongly reactive to self-antigens.

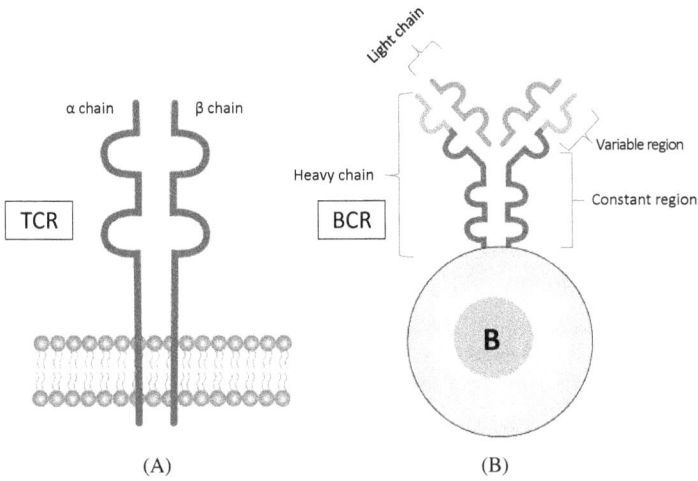

Figure 1.17 **(A) TCR on T cell.** TCR consists of two chains, namely, α and β chains. Variable region binds to antigen-MHC complex. **(B) BCR on B cell.** BCR is a membrane-bound antibody molecule.

1.13 T cell receptor

TCRs are composed of two chains referred to as alpha and beta chains. Each chain has a variable region, constant region, hinge, transmembrane and cytoplasmic domain (Figure 1.17). Different TCRs have different specificities and this is conferred by the variable region/antigen binding site which is different for every T cell clone. Once the TCR recognises specific antigens bound to MHC, the T cell is activated. Kinases associated with TCR/CD3 complex are then activated resulting in activation of gene expression, and the T cell proliferates and carries out its effector functions.

1.14 B cell receptor

The B cell receptor (BCR) is a membrane-bound antibody molecule. When an antigen binds to a BCR, accessory molecules associated with the BCR transmit signals into the B cell.

When the TCR or BCR binds to its specific antigen, the T and B cells are activated respectively. Each T and B cell will recognise only one particular antigen. The cells proliferate and divide repeatedly resulting in production of clones of cells, i.e. cells with identical TCRs or BCRs. This expansion is driven by IL-2.

Some of these cells that have encountered antigens will become effector cells and some become memory cells.

Nutrients provide a supply of nucleotides (for DNA and RNA synthesis), amino acids (for protein synthesis), fatty acids, bases, phosphates (for phospholipid synthesis) and other lipids (e.g. cholesterol) for generating these immune cells.

Defence against intracellular pathogens requires CMI since the pathogens are not accessible to antibodies when they are inside the infected cell.

There are two major populations of T cells that mediate CMI. They are T helper and cytotoxic T cells (Figure 1.18). T helper cells are usually $CD4^+$ T cells and they help B cells become activated. Cytotoxic T cells are usually $CD8^+$ T cells and they kill infected or cancerous cells.

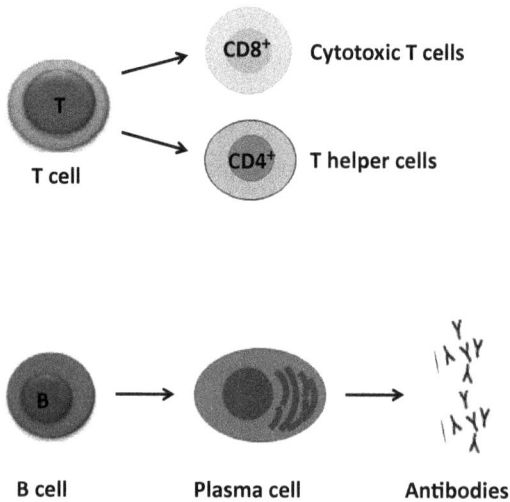

Figure 1.18   Cells of the adaptive immune system. T cells are classified as T helper or cytotoxic T cells which mediate CMI. B cells produce antibodies which mediate humoral immunity.

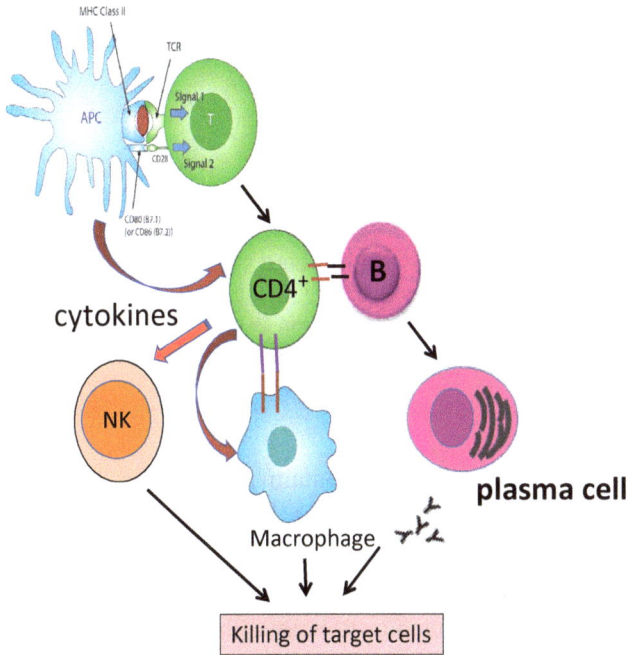

Figure 1.19   Interactive network of cells of the immune system. Antigen presentation to TCR of CD4$^+$ by APC results in CD4 T helper cells, which interact with B cells to elicit formation of plasma cells which produces antibodies. CD4$^+$ T helper cells also activate macrophage and NK cells via secretion of cytokines. Macrophage and NK cell activation leads to killing of tumour or virally infected cells. Antibodies which are recognised by antigen on tumour or infected cells will facilitate killing by NK and macrophage cells via antibody-dependent cytotoxicity.

1.15 The functions of CD4$^+$ T helper cells are to

(1) interact with B cells to activate them — produce antibodies
(2) produce cytokines to activate T and B cells and macrophages to clear infection. The type of cytokines produced depends on the class of infectious agent (Figure 1.19).

1.16 The functions of cytotoxic T cells

Virally infected cells will display viral antigens on MHC class I molecule. Cytotoxic T cells are activated when it recognises antigenic peptide asso-

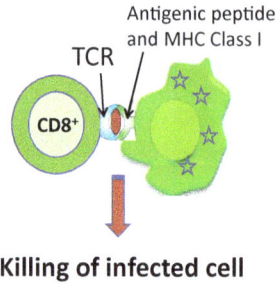

Killing of infected cell

Figure 1.20   Recognition of antigenic peptide in association with MHC by TCR on CD8$^+$ CTL triggers activation of CD8$^+$ CTL and results in killing of infected cell.

ciated with MHC class I. They contain granules for killing which are released once activated (Figure 1.20). These granules include perforin which forms pores in membrane and granzymes which are proteolytic enzymes. They destroy structural cytoskeleton proteins and degrade DNA, resulting in killing of infected or tumour cells.

1.17 B lymphocytes and antibody production

B cells recognise a single antigen via the BCR. Once the B cell is activated, it differentiates into plasma cells to produce antibodies that will recognise the specific antigen that was recognised by the B cells to induce antibody production (Figure 1.21).

1.18 Antibodies

Antibodies are proteins produced by differentiated B cells known as plasma cells. They are important in neutralisation of pathogens such as bacterial toxins and viruses (Figure 1.22). Antibodies recognise specific antigens. Once antigen is recognised, pathogen can be neutralised, clumped, agglutinated or phagocytosed by macrophages. There are five classes of human antibodies (isotypes), namely, IgG, IgM, IgA, IgD and

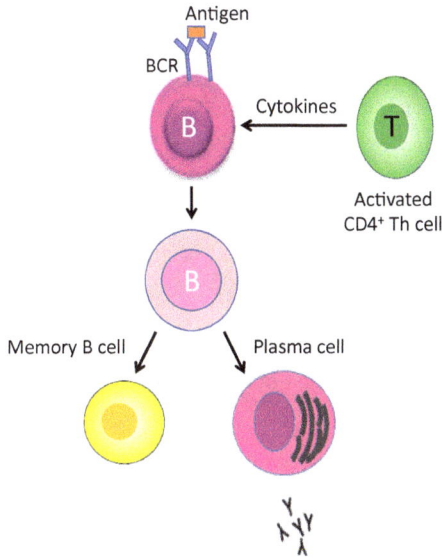

Figure 1.21   Steps involved in B cell activation and generation of plasma and memory cells.

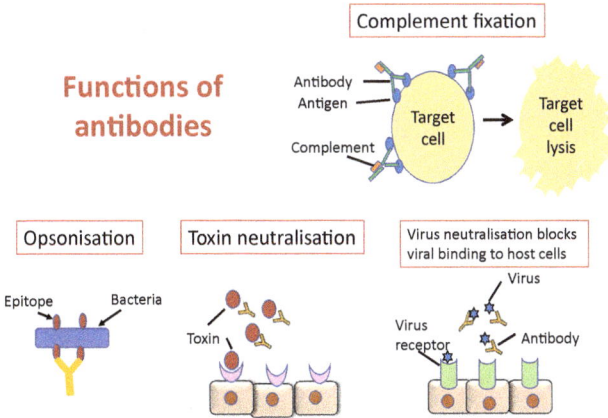

Figure 1.22   Functions of antibodies.

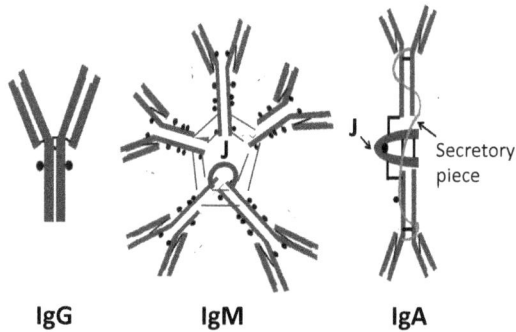

Figure 1.23   Schematic representation of the structure of antibody isotypes IgG, IgM and secretory IgA.

IgE. IgM and IgD are monomeric in structure which is similar to IgG (Figure 1.23). IgG is the most abundant antibody in blood. It can cross the placenta and confer protective immunity for the first six months of an infant's life. IgM is pentameric in structure (Figure 1.23) and is produced by plasma cells during the primary immune response. Secretory IgA is dimeric in structure (Figure 1.23) and helps to prevent attachment of pathogens to mucosal surfaces such as those in the respiratory, gastrointestinal and urinogenital tracts. IgD is attached to the surface of B cells. IgE binds to mast cells and basophils, causing the release of histamine when these cells are activated.

1.19 Gastrointestinal immune system

1.19.1 *Organisation and function of the gastrointestinal immune system*

The gastrointestinal tract has a highly organised mucosal immune system that plays a critical role in immunosurveillance and homeostasis in the gut. The immunological components of the gut play a key role in providing protective immunity against pathogenic microorganisms, and tolerating food substances and commensal bacteria. Thus, the gut immune system serves as an essential barrier to prevent invasion by pathogens and also to prevent the development of intestinal immune diseases such as inflammatory bowel diseases and food allergies. Deficiency in nutritional intake has been shown to increase susceptibility to infections and increase risk of allergic and inflammatory diseases. In addition, nutritional components such as dietary or *de novo* synthesised vitamins and lipids play an important role in the development, maintenance and regulation of the gut immune responses.

The lymphoid tissue within the gastrointestinal tract is referred to as the gut-associated lymphoid tissue or GALT. Several subsets of intestinal epithelial cells including goblet cells, Paneth cells, microfold (M) cells, enteroendocrine cells, and columnar epithelial cells are located at the epithelial layer of the gastrointestinal tract (Figure 1.24). As summarised in the table below, mucus is secreted by globlet cells, anti-microbial peptides such as α-defensins are produced by Paneth cells, and β-defensin are produced by other endothelial cells. Proteins in the form of mucins cover the mucosa to create a physical barrier to pathogens. Mucin contains high concentrations of defensins and other antibacterial molecules such as lactoferrin and lysozyme. Adherence of bacteria to epithelial cells is impaired by lactoferrin. Bacteria is killed by lysozyme. All of these components are important as innate defences and work together with the adaptive immune system that produces secretory IgA (sIgA), which prevents the transmigration of the invading pathogen from lumen into the internal environment.

Under conditions of severe stress such as in critical illness, prolonged use of antibiotics, prolonged GI tract starvation, and with alterations in normal oral intake, the mucosal defences are weakened due to reduced mucin and reduced antimicrobial peptides, resulting in increased vulnerability to pathogen invasion.

GALT including Peyer's patches, isolated lymphoid follicles, and colonic patches are also referred to as the inductive sites. The lymphocytes in GALT are distributed in three basic populations, namely, Peyer's patches, lamina propria lymphocytes and intraepithelial lymphocytes. One important component of the gastrointestinal immune system is the M cell. They are located in the intestinal epithelium over the Peyer's patches. Dendritic cells within the Peyer's patches endocytose antigens and transport them into the T cell region, and subsequently germinal centers in the

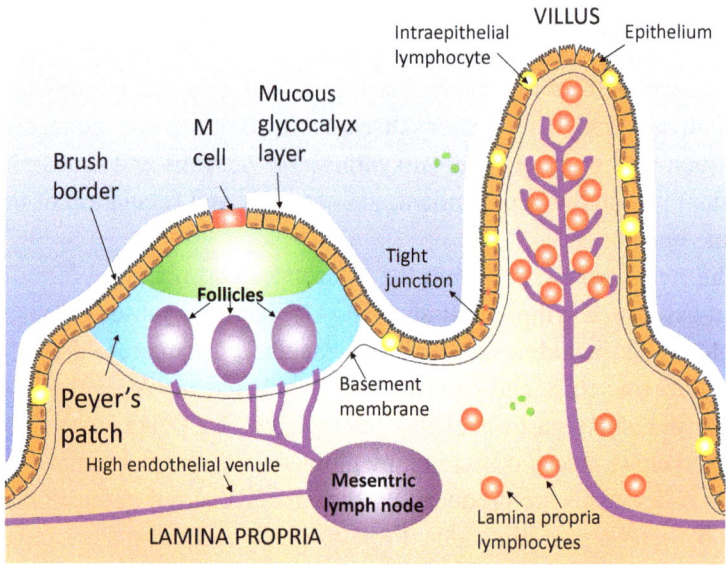

Figure 1.24   Organisation of the gastrointestinal system.

Key points on the functions of components of the gastrointestinal system:

- Intestinal epithelial cells: monolayer of cells covering luminal surface of the intestine. Serves as a barrier against invasion by pathogens and is responsible for absorption of nutrients. Cells include globlet cells, Paneth cells, enteroendocrine and columnar epithelial cells.

- M cells: found in the epithelium of Peyer's patches; transport antigens from the lumen to the lymphoid follicles.

- Peyer's patches: group of lymphoid nodules at the intestinal wall. Consists of the dome area, B cell follicles and interfollicular T cell areas.

- Lamina propria: connective tissue between intestinal epithelium and thin layer of gastrointestinal muscle tissue.

- Mesenteric lymph nodes: lymph nodes located at the base of mesentery. They collect lymph draining from the intestinal mucosa.

GALT or via draining lymphatics into the mesenteric lymph nodes, for the initiation of T and B cell responses. At the mesenteric lymph nodes, the T and B cells mature and/or proliferate and further differentiate into effector cells such as IgA-producing plasma cells, cytotoxic T cells, regulatory T cells (Tregs), and Th17 cells.

Effector cells traffic out of the lymph node and join the blood circulation via the thoracic duct to reach the submucosal sites in the upper and lower respiratory tracts as well as the small and large intestines. In these sites, B lymphocytes differentiate into plasma cells capable of producing IgM and IgA. Dimeric IgA secreted from plasma cells binds to molecules of polyimmune globulin receptor located on the basal surfaces of the mucosal cell and is transported to the luminal surface. IgA complexed to a small segment of pIgR (secretory component), referred to as sIgA, is then released into the lumen.

sIgA plays a key role in mucosal immunity by binding to bacterial surface antigens to block attachment of the bacteria to the mucosa. These immunological networks in the gastrointestinal immune system are important to orchestrate both active and quiescent immune responses for immunosurveillance and homeostasis in the gut.

1.19.2 *Gastrointestinal immune system and nutrition*

Cells of the immune system utilise glucose, amino acids and fatty acids as fuels for energy (ATP) generation. ATP generation via the mitochondrial respiratory chain involves electron carriers and a range of coenzymes which are usually derivatives of vitamins and have Fe or Cu at their active site. Recognition of antigens by cells of the innate immune system induces production of acute phase proteins, cytokines and lipid-mediators prostaglandins and leukotrienes. Induction of the innate responses drives the activation of the adaptive immune response which involves cellular proliferation. Proliferation of immune cells requires synthesis of DNA and cellular components. Cellular proliferation is a key component of the immune response, and involves replication of DNA and cellular components (proteins, membranes, intracellular organelles) as well as a supply of nucleotides, amino acids, fatty acids, bases, phosphates (for phospholipid synthesis) and other lipids (e.g. cholesterol). Dietary components including fatty acids, amino acids and minerals are therefore essential. Micronutrients such as folate, iron, zinc and magnesium are also involved in nucleic acid synthesis. Vitamins, such as vitamins A and D, play a key role in immune function by enhancing maturation of APCs and differentiation of naïve T cells. Thus, nutrients play an essential role in immune function and maintain a healthy immune system.

1.20 Antigen recognition, lymphocyte activation and differentiation

1.20.1 *What are the antigens that are recognised by the immune system?*

The key to a healthy immune system is that it has the ability to distinguish between self and non-self cells and proteins. Anything that can activate an immune response is referred to as an antigen. Examples of the types of antigens (including proteins, DNA, and lipids) which are recognised by the immune system are shown in Figure 1.25. Examples of viral proteins that are used for vaccine development include the hepatitis B surface antigen, influenza haemagglutinin and papillomavirus E6 and E7 antigens. An example of an antigen from bacteria is tetanus toxin from *Clostridium tetani*. It can be detoxified to produce tetanus toxoid, which is used as the vaccine antigen to prevent tetanus.

1.20.2 *How does the immune system distinguish between self versus non-self antigens?*

As our immune system matures, it learns not to react to self-molecules and this is referred to as tolerance. The immune system is also tolerant to ingested food substances. In contrast, cell surface receptors such as TCRs and BCRs can specifically recognise antigens from pathogens such as

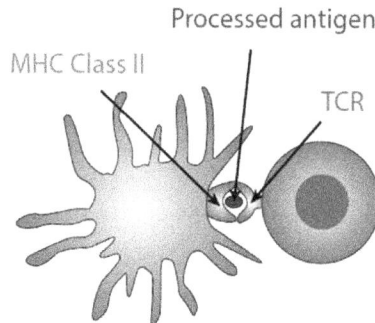

Figure 1.25 Step 1 in T cell activation (Signal 1). The TCR recognises antigenic peptide in association with MHC presented by the APC e.g. a dendritic cell.

bacteria. Once these non-self molecules are recognised, an immune response is mounted. Self-molecules such as DNA and nucleoproteins which are normally hidden from the immune system but are released by injured or dead cells can serve as antigens, causing the immune system to attack its own cells. An example is immune attack on pancreatic islet cells in type I diabetes mellitus. In some individuals, harmless environmental antigens can trigger immune responses that gives rise to allergy. Cancer can also arise when the immune response is defective and mutated cells are not eliminated, and/or when T cells are dormant.

1.20.3 *Antigen processing and presentation*

Dendritic cells phagocytose pathogens, process antigens into peptides and present the antigenic peptides in association with the MHC on the surface of the APC such as a dendritic cell (Figure 1.25). Antigens captured by the APC are degraded and loaded onto the MHC Class II, whilst antigens synthesised in the APC are degraded and presented in association with MHC Class I (Figure 1.26). Naïve T cells are activated by coming in contact with a dendritic cell which presents a peptide/MHC complex to the TCR. This process occurs in secondary lymphoid tissues such as lymph nodes.

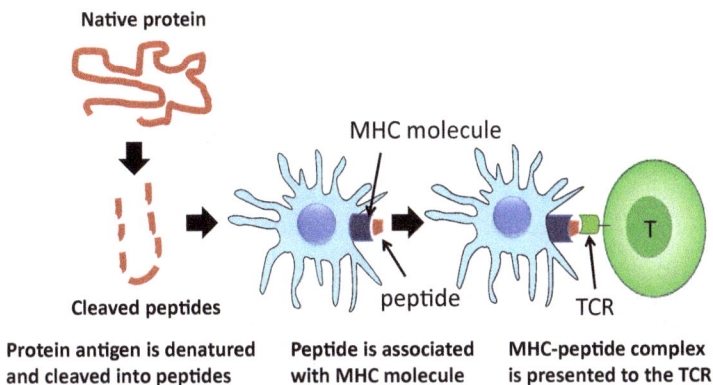

Native protein

MHC molecule

Cleaved peptides

peptide

TCR

Protein antigen is denatured and cleaved into peptides   **Peptide is associated with MHC molecule**   **MHC-peptide complex is presented to the TCR**

Figure 1.26   Steps in antigen processing and association with MHC for presentation to TCR.

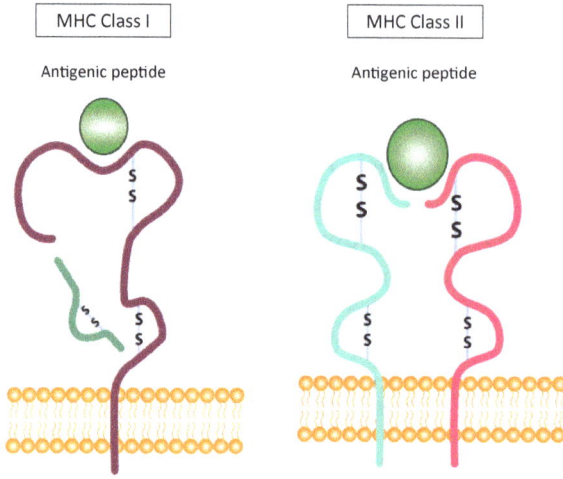

Figure 1.27   Schematic representation of MHC Class I and II proteins.

1.20.4 *What is Major Histocompatibility Complex molecule?*

There are two MHC proteins. Class I MHC proteins are found on all cells except erythrocytes, and Class II MHC proteins are found on certain cells of the immune response. They are encoded by genes of the MHC on chromosome 6 and are unique to an individual (Figure 1.27). Class I MHC proteins consist of HLA-A, B and C. Class II MHC proteins are HLA-DQ, DP and DR (Figure 1.27).

Each MHC molecule has a deep groove that displays a peptide derived from processed antigen (Figure 1.27). MHC proteins on the surface of the APCs bind to peptides of foreign antigens which are recognised by the TCR.

1.20.5 *T cell activation and co-stimulation*

For a naïve cell to become an effector cell, it needs to receive signals by cell-to-cell contact with APCs such as the dendritic cells in secondary lymphoid tissues and from cytokines in the immediate microenvironment. When TCR of naïve T cells binds to processed antigen in the form of a peptide in association with an MHC molecule on the APC, this provides signal 1 for T cell activation. TCR binds to the antigenic peptide linked to MHC Class II protein (Figure 1.28) whilst for cytotoxic T cells, the antigenic peptide is complexed to MHC Class I protein. The other T cell surface proteins such as CD4 and CD8 help to maintain coupling during antigen recognition.

Before a T cell can undergo clonal expansion, it must bind one or more co-stimulatory molecules (Figure 1.29). Expression of co-stimulatory

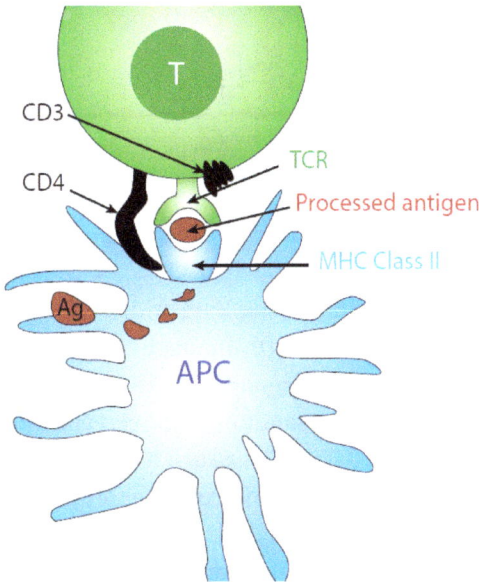

Figure 1.28 Processed antigen is linked to MHC Class II and presented to TCR of CD4^{+} T cell.

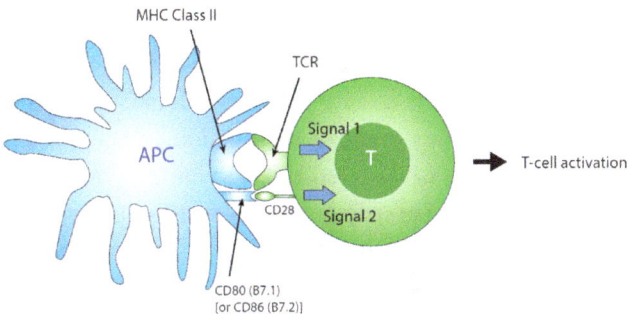

Figure 1.29 Molecules involved in signals 1 and 2 for T cell activation.

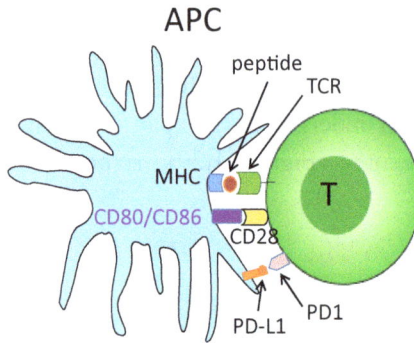

Figure 1.30   Co-stimulatory and co-inhibitory molecules and their partners on APC and T cells.

receptors on APCs are up-regulated when APCs are activated. CD80 (B7.1) or CD86 (B7.2) binding to CD28 on the surface of T cells are crucial as co-stimulatory signals (Figure 1.30). Interaction of CD80/CD86 on APC with CD28 on T cells provides signal 2 for T cell activation. Without co-stimulation, T cells become tolerant to the antigen and are unable to divide as well as secrete cytokines. Cytokines such as IL-2 provide signal 3 for T cell activation and proliferation. The T cells that are activated enlarge, proliferate and form clones. They also differentiate into various T helper subsets and perform their effector function. Primary T cell responses peak within a week after exposure to antigen. The T cells then undergo apoptosis (programmed cell death) between 7–30 days. The effector T cell activity diminishes as the amount of antigen declines. When the invading pathogen is eliminated, T cell activation stops. This inhibition is provided by inhibitory molecules such as CTLA-4 which bind to CD80/CD86 (B7.1/B7.2) (Figure 1.31). The disposal of activated effector T cells prevents tissue damage and is a protective mechanism for the body. Memory T cells remain and mediate secondary responses to the same function.

Thus, T cell function is initiated through antigen recognition by the TCR and is regulated by a balance between co-stimulatory and inhibitory signals (also referred to as immune checkpoints). Immune checkpoints are crucial for protecting tissue damage when immune system responds to pathogens, and also for maintenance of self-tolerance which prevents autoimmunity.

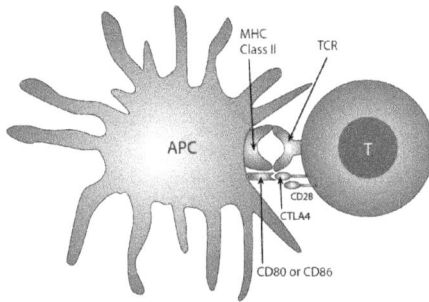

Figure 1.31 Binding of CTLA-4 to CD80/D86 (B7.1/B7.2) attenuates T cell activity.

1.20.6 *T cell differentiation*

For a naïve T cell to become an effector T cell, it must receive signals by interacting with other cells including dendritic cells, macrophages, B cells as well as cytokines. Activated naïve CD4$^+$ T helper cells undergo differentiation to different helper T cells such as Th1, Th2, Th17 and Tregs (Figure 1.32). The cytokines in the environment of the T cells and dendritic cells drive these differentiation pathways.

When naïve T cells are activated through the TCR in a microenvironment which is abundant in IL-12 (secreted by activated dendritic cells and macrophages in response to infection and inflammation) and IFN-γ (produced by natural killer cells and T cells), they induce the expression of the transcription factor TBET, which drives differentiation to Th1 cells (Figure 1.32). Th1 cells produce IFN-γ, which is known as their 'signature cytokine' and is important in activating macrophages. Th1 cells mediate immune response against intracellular pathogens.

When naïve CD4$^+$ T cells are in a microenvironment abundant in IL-4, expression of transcription factor GATA3 is induced and this drives the differentiation pathway to Th2 cells. Th2 cells produce IL-4 as their 'signature cytokine' along with IL-5 and IL-13. They mediate host defence against extracellular parasites such as helminth parasites (Figure 1.33), and are also important in induction and persistence of allergic diseases such as asthma. IL-4 induces B cell antibody class switching to IgE, which is important in defence against parasitic infections. Th2 responses are predominant in allergy states such as asthma.

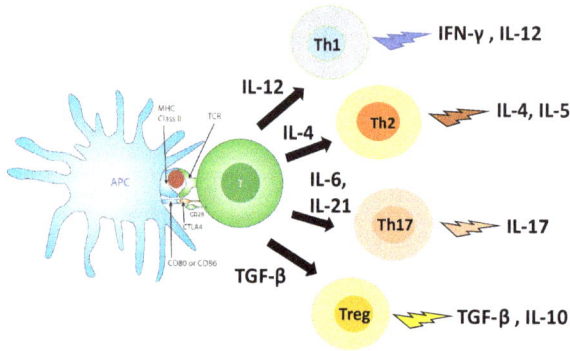

Figure 1.32   Differentiation pathways of naïve T cells.

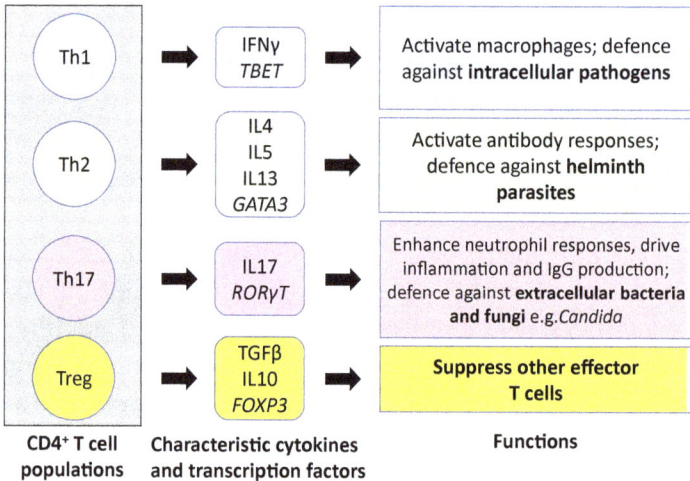

Figure 1.33   Signature cytokines and transcription factor of T helper cells and their functions.

A third Th subset is Th17. IL-6 and TGF-β promote differentiation to mouse Th17 cells. Differentiation of human Th17 cells require IL-6 and IL-21. Th17 cells secrete IL-17 and IL-21, which induce secretion of CXCL8 and other cytokines, leading to recruitment of neutrophils to the site of infection and production of antimicrobial peptides. Expression of the transcription factor RORγT is required to switch on the expression of

IL-17. Th17 cells are abundant in the skin and gut mucosa, where they function as effector cells against pathogens such as *Candida albicans* and *Staphylococcus aureus*. Th17 cells also play an important role in the pathology of autoimmune and allergic diseases (Figure 1.33).

Naïve T cells activated in the presence of TGF-β but in the absence of IL-6 and other proinflammatory cytokines become induced Tregs. Tregs play a central role in immune regulation. There are at least two types of Tregs, namely, naturally occurring Tregs in the thymus, which are responsible for clonal deletion of self-antigens, and inducible Tregs, which are in the periphery and are inducible from naïve T cells.

Tregs express the transcription factor FOXP3 and secrete TGF-β and IL-10, which modulate inflammatory responses (Figure 1.33). Induced Tregs are Treg cell populations that arise from precursor cells in the periphery, as opposed to naturally occurring Tregs that originate from the thymus. Inducible Tregs can be found on the mucosal surfaces of the intestine. Tregs play an important role in shutting down immune responses after successful elimination of pathogens, limiting tissue damage and facilitating healing. Tregs can also prevent autoimmunity and induce immune tolerance to food antigens.

1.20.7 *B cell activation*

Antigen binding to BCR triggers cross-linking of BCR and phagocytosis of antigen. Upregulation of co-stimulatory molecules such as CD40 and interleukin receptors such as IL-2 receptors on B cells also occurs (Figure 1.34). The internalised antigen is processed into peptides and presented to the TCR. Binding of peptide to TCR and interaction of CD40 ligand with CD40 on B cells activates T cells to produce cytokines, which provides help for antibody production and induces antibody isotype switching.

1.21 Vitamins such as vitamins A and D can influence T cell differentiation

Retinoic acid, a metabolite of vitamin A, increases IL-4 production and inhibits differentiation to Th17. Retinoic acid in combination with TGF-β induces naïve T cell differentiation to inducible Tregs.

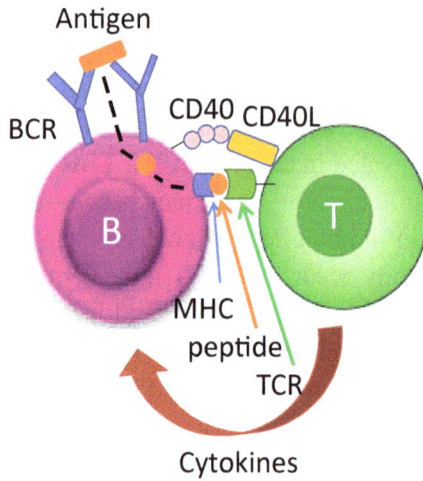

Figure 1.34   B cell activation. Antigen binding to BCR is internalised, processed into peptide and presented to TCR in association with MHC. T helper cells which recognise the antigen provide co-stimulation by expressing CD40L. Interaction between CD40L, CD40 and cytokines produced by Th cells induce antibody isotype switching and somatic hypermutation.

Vitamin D_3 stimulates production of IL-10 and increases number of Tregs. It also suppresses production of IFN-γ, IL-17, IL-21 and IL-22, which in turn suppress proliferation of Th1 and Th17 cells, which have been reported to support pro-inflammatory responses.

1.22   Nutrient imbalance and immune function

Nutrient imbalance can contribute to immune suppression and this is observed in states of malnutrition. Nutritional balance is essential for the development of cells and organs of the immune system. Overnutrition can lead to obesity and diseases linked to metabolic syndromes (a cluster of conditions which includes high blood sugar, hypertension, excess body fat around the waist, and abnormal cholesterol and triglyceride levels that when occurred together can increase risk of heart disease, stroke and type II diabetes). In primary malnutrition states, atrophy of lymphoid organs and loss of immune function lead to increased susceptibility to infection. Nutrient needs change during aging, and gradual loss of immune functions in the elderly are associated with thymic involution and accumulated antigenic exposures. Nutrient supplementation should be beneficial in the elderly, but it is complicated as it can depend on the health status of the individual, while there is also a lack of biomarkers to assess the benefits of supplementation. The immunological consequences of vitamin and mineral deficiencies and overdosing are covered in their specific chapters.

Summary of Chapter 1

The immune system consists of a network of lymphatic system, organs, tissues, cells and proteins that cooperate to defend the body against foreign pathogens while tolerating self-antigens or non-harmful substances such as food. Antigens are transported from the site of infection to the lymph nodes and spleen via the lymphatic system. Effector cells and antibodies exit the lymph node via the efferent lymphatic vessels into the blood circulation. The lymph nodes and spleen are referred to as the secondary lymphoid organs. Cytokines and adhesion molecules play key roles in cell-to-cell communication. Cells of the immune system originate from the bone marrow.

The three lines of defence of the immune system consist of the physical barriers (e.g. skin, gastrointestinal and respiratory mucosa), chemical barriers (e.g. enzymes and chemicals from skin, stomach and mucosa surfaces) and cells and proteins of the innate and adaptive immune system. The characteristic features of neutrophils, macrophages, eosinophils, mast cells, basophils, natural killer cells, T and B lymphocytes are described. The gastrointestinal tract mucosal surface is exposed to invasive pathogens, and it is thus important that the immune system is established at this location. Paneth cells are specialised epithelial cells of the small intestine and they secrete a number of antimicrobial substances such as defensins, lysozyme and phospholipase A2 to kill invading bacteria, fungi and certain viruses. The gastrointestinal tract is a lymphoid organ, and lymphoid tissues within it are referred to as the gut-associated lymphoid tissue or GALT. The number of lymphocytes is approximately equal to that in the spleen and they are distributed in three populations, namely, the Peyers patches, lamina propria and intraepithelial regions. The Peyer's patches are similar to lymph nodes and they are located in the mucosa extending to the submucosa of the small intestine. The lamina propria lymphocytes are scattered in the lamina propria and majority of the cells are IgA-secreting B cells. Intraepithelial lymphocytes are located in the epithelium between luminal epithelial cells beneath the tight junctions. Another important component is the M cell which is located in the intestinal epithelium and plays a key role in transporting proteins and peptide antigens into the tissue where they are taken up by the dendritic cells and macrophages. These APCs present the processed antigens to T cells in T cell areas and subsequently to germinal centers in the GALT for the initiation of T and B cell responses. The sensitised T and B cells then migrate to the mesenteric lymph nodes to mature and/or proliferate and further differentiate into effector cells such as IgA-producing plasma cells, cytotoxic T cells, Tregs, and Th17 cells prior to their release into the thoracic duct. Foreign antigens are taken up by APCs, processed and presented to the TCR in association with the MHC molecule (Signal 1). This is a key step in driving the adaptive immune response. Activation of the T cells also require the interaction of co-stimulatory molecules, CD28 on T cells and CD80/CD86 on the APCs (Signal 2). Cytokines produced in the activation of naïve T cells drive differentiation of naïve T cells to various T helper subsets such as Th1, Th2, Th17 or Tregs. Retinoic acid (a Vitamin A metabolite) and Vitamin D_3 (active form of Vitamin D) regulate the differentiation of naïve T cells. B cell activation involves the interaction of the antigen binding to the BCR, receptor cross-linking and interaction of CD40 and CD40 ligand, resulting in antibody production and antibody isotype switching. Nutritional imbalance can cause immunosuppression as evidenced from malnutrition states. In primary malnutrition states, atrophy of lymphoid organs and loss of immune function lead to increased susceptibility to infection. Overnutrition can lead to obesity and diseases linked to metabolic syndrome. Nutrient needs change during aging, and gradual loss immune functions in the elderly are associated with thymic involution and accumulated antigenic exposures.

Chapter 2

Vitamins

Learning objectives

After studying this chapter, you should be able to:

1. Understand basics in metabolism of vitamins A, B, C, D and E
2. Explain the effects of these vitamins on immune cells
3. Understand the consequences of deficiency in these vitamins on human health

2.1 Introduction

Micronutrients which include vitamin A, vitamin C, vitamin B_{12} (cobalamin), vitamin B_6 (pyridoxine), vitamin B_9 (folate), vitamin D, vitamin E, zinc, copper, iron, and selenium are required for the immune system to function effectively. They can contribute to immune defenses at three levels, namely, by supporting physical barriers (skin/mucosa), promoting innate immunity and cell-mediated immunity, and antibody production. Most of the effects of micronutrients described in this chapter are from *in vitro* and animal model studies. The immunological consequences of micronutrient deficiencies in humans provide information on their essential role in protecting the host from susceptibility to infections.

Vitamins A, C, E, and zinc can enhance the epithelial barrier function. Vitamins A, C, B_{12} (cobalamin), B_6 (pyridoxine), B_9 (folate), D, and E, and minerals iron, zinc, copper, and selenium have synergistic effects on immune cells. Finally, all these micronutrients, except vitamin C and iron, are essential for antibody production.

Vitamins and minerals are essential for immune function as inadequate intake of these vitamins and minerals may lead to depression of immunity, which increases risk of susceptibility to infections and aggravates malnutrition. Recent discoveries on the link between inflammation and chronic disease have driven interest in vitamin and mineral supplementation for prevention of chronic diseases. Oxidative stress due to excessive inflammation and low levels of antioxidants is linked to the aetiology of chronic diseases. Chronic inflammation leads to destruction of host tissues by the activated immune system. Thus, the immune system needs to be tightly regulated. Modulation of immune function by nutritional means appears to be a rational approach in treating or preventing chronic inflammatory conditions. Immunomodulatory effects of vitamins A and D and their metabolites, or antioxidants such as vitamins and minerals, implicate their potential usefulness.

2.2 Vitamin A

2.2.1 *Introduction*

Vitamin A consists of three essential fat-soluble molecules: retinol, retinal and retinoic acid. Retinol is the form in which vitamin A is stored. Retinal is crucial for vision. Retinoic acid is an active metabolite of vitamin A and functions like a hormone by binding to two nuclear receptors, retinoic acid receptor (RAR) and retinoid X receptor (RXR) which regulates gene transcription.

2.2.2 *Dietary sources of vitamin A*

Vitamin A is obtained from the diet either as *all-trans*-retinol, retinyl esters or β-carotene as it cannot be synthesised by the body. It must be absorbed by the intestine from the diet. Vitamin A is found as retinol in foods of animal origin such as fish, liver, milk, butter, cheese, and egg yolk (Table 2.1).

Table 2.1   Examples of common food sources of vitamins.

Vitamin	Examples of Common Food Sources
A	Carrot, pumpkin, liver, cantaloupe, fish, milk, butter, cheese, egg yolk
B_1 (thiamine)	Whole grains, tuna, black beans, green peas, pork
B_2 (riboflavin)	Whole grains, oatmeal, eggs, mushroom, spinach, pork
B_3 (niacin)	Whole grains, tuna, salmon, beef, chicken, mushrooms
B_6 (pyridoxine)	Tuna, turkey, beef, chicken, whole grain, banana, chick peas
B_9 (folate)	Avocados, dark green leafy vegetables, broccoli, liver, brewer's yeast, oranges, lentils
B_{12} (cobalamin)	Crab, beef, salmon, soy milk, yogurt, egg, cottage cheese
C	Broccoli, blackcurrants, kiwi fruit, strawberries, citrus fruits such as orange and tangerine, papaya
D	Oily fish, eggs, milk, cereals
E	Nuts, wheat germ, whole grains, eggs, fish, avocado, vegetable oils

Vitamin A does not occur in plants, but many plants contain carotenoids such as beta-carotene that can be converted to vitamin A in the intestine and other tissues. Plant sources of vitamin A are orange- and yellow-coloured fruits such as papaya and mango, and vegetables such as carrots, yellow sweet potatoes, and green leafy vegetables (spinach). These plant sources of carotenoids are converted to retinol.

2.2.3 *Vitamin A metabolism*

As mentioned, vitamin A is obtained from the diet either as *all-trans*-retinol from animal sources or β-carotene from plants. β-carotene is converted into retinal by oxidation and then reduced to retinol (Figure 2.1). Retinol from the diet is transported in the blood to the liver and other tissues. The liver is the site for esterification of retinol to retinyl esters and storage. In other tissues, including gut-associated immune cells, retinol can be converted to retinoic acid by two enzymatic steps. The first step involves oxidation of retinol to retinal by alcohol dehydrogenase (ADH), followed by conversion to retinoic acid by retinaldehyde dehydrogenase (RALDH). RALDH is found in gut-associated cells including dendritic cells (DCs) in Peyer's patches, intestinal epithelial cells and mesenteric lymph nodes.

Retinyl Esters

| Retinyl ester hydrolase | | Lecithin retinol acyltransferase |

Retinol — Mobilisation to extrahepatic tissue

Alcohol dehydrogenase (ADH)

β-carotene

Retinal ← β-carotene 15, 15′ monooxygenase

Retinaldehyde dehydrogenase (RALDH)

Retinoic acid

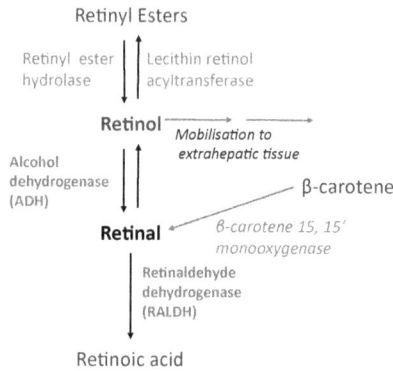

Figure 2.1   Metabolism of retinol and β-carotene. β-carotene is converted into retinal by oxidation. Retinol obtained from diet is converted to retinal by alcohol dehydrogenase (ADH) or retinol dehydrogenase. Retinal is then converted to retinoic acid by retinaldehyde dehydrogenase (RALDH).

Immune effects of retinoic acid are elicited when it binds to the retinoic acid receptor on immune cells. Catabolism of retinoic acid occurs in the liver and other tissues, and metabolites are eliminated in the urine and bile.

2.2.4 *Functions of vitamin A*

Vitamin A is important for energy homeostasis. Retinol is essential for generation of ATP in the mitochondria. When cells are deprived of retinol, mitochondrial respiration and ATP synthesis fall. When retinol is restored to physiological concentration, the cells recover.

Vitamin A was initially coined "the anti-infective vitamin" because it plays an essential role in the normal function of the immune system. Vitamin A is required for innate and adaptive immunity. The skin and mucosal cells lining the gastrointestinal, respiratory and urinogenital tracts act as a barrier and form the first line of defense against infection. Deficiency of vitamin A impairs mucosal and epithelial barrier functions, changes immune responses and increases vulnerability to infections. Entry of pathogens can occur when mucus secretion from mucosal surfaces is impaired, and when epithelial barriers such as cilia from respiratory epithelium and microvilli in the gastrointestinal tract are lost. Vitamin A also potentiates the antibody response especially IgA at the mucosa, and maintains and restores

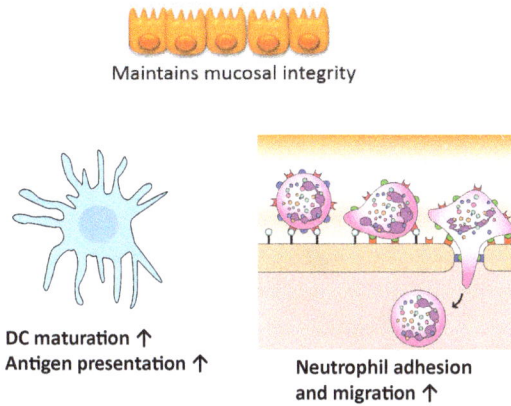

Figure 2.2   Role of vitamin A in innate immunity. Vitamin A maintains mucosal integrity which functions as the body's first line of defense. It enhances DC maturation and antigen presentation, and increases neutrophil adhesion and migration to the site of infection.

the integrity and function of all mucosal surfaces (Figure 2.2). Secretory IgA at the gut lining provides protection against harmful pathogens. Antibody responses are also impaired in vitamin A deficiency.

Key points on consequences of vitamin A deficiency:

- Impairment of mucosal and epithelial function
- Impairment of antibody responses
- Impaired neutrophil chemotaxis, phagocytosis and intracellular killing
- Reduced natural killer (NK) cell numbers
- Shift to Th2 cell differentiation
- Increased HIV pathogenesis, increased susceptibility to respiratory infections, diarrhoea and severe measles in children, which is likely due to impairment of mucosal integrity and reduced antibody production as a result of Th2 response impairment

---

Retinoic acid, derived from vitamin A in the diet, promotes IgA production. It is secreted by antigen-presenting cells (APCs), including macrophages and DCs at mucosal interfaces and associated lymph nodes. In brief, it performs the following functions: (1) maintains mucosal integrity, (2) enhances antigen-presenting capacity of DCs, and enhances migration to draining lymph node, (3) increases neutrophil adhesion and migration, (4) induces differentiation of naïve CD4 T-lymphocytes into Th2 or

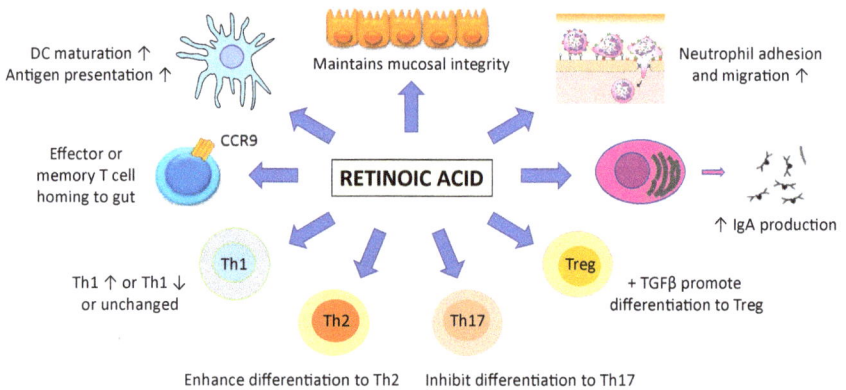

Figure 2.3   Summary of the effects of retinoic acid (an active metabolite of vitamin A) on cells of the immune system.

induced regulatory T-lymphocytes (iTregs), but at high concentrations inhibits differentiation to Th17 cells, (5) promotes the differentiation of naïve CD4$^+$ T cells into Th1 cells and induces proinflammatory cytokine production by effector T cells in response to infection (but some studies show opposite effects or no change) (6) induces IgA class switching, and (7) induces homing of lymphocytes to the gut (Figure 2.3).

2.2.5 *Mechanism of action of retinoic acid on immune cells*

1. **Enhanced DC maturation and antigen presenting capacity**

   The presence of inflammatory cytokines, such as tumour necrosis factor (TNF), allows enhancement of maturation of DCs and antigen presenting ability by retinoic acid, leading to induction of T cell proliferation and activation.

2. **Enhanced DC migration via upregulation of matrix metalloproteinase 9 expression**

   Matrix metalloproteinase (MMP)-9 degrades the extracellular matrix by cleaving gelatin, collagen types I and IV, and laminin. In inflammation, DCs require MMP-9 activity to migrate and pass through tight epithelial junctions. This increased migratory activity of DCs has been shown *in vitro* by culturing of immature DCs with retinoic acid and in animal tumour models. Thus, increased expression of MMP9 by retinoic acid can increase the migration of tumour-infiltrating DCs to the draining lymph node, which can then boost tumour-specific T cells.

3. **Enhanced T cell proliferation and differentiation of T cells to Th2 or regulatory T cells**

   Retinoic acid also plays a role in T cell proliferation and differentiation (Figure 2.4). Retinoic acid can enhance T cell proliferation via upregulation of IL-2 secretion and T cell signaling.

In vitro and animal studies show that retinoic acid promotes differentiation to Th2 cells by inducing expression of IL-4 and Th2-promoting transcription factor GATA3. In a mouse model of asthma, vitamin A supplementation increased Th2 cytokine expression and disease severity. In contrast, vitamin A deficiency had an opposite effect. These findings suggest that vitamin A promotes differentiation to Th2 cells.

Th17 cells function as effector cells against intracellular pathogens and are abundant at mucosal surfaces. Th17 cells play a major role in inflammatory response and has an important role in the pathology of

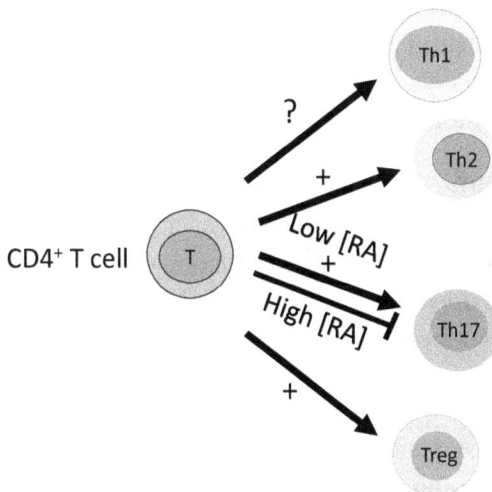

Figure 2.4   Retinoic acid influences differentiation of T helper cells.

autoimmune and allergic diseases. TGF-β and IL-6 induce the differentiation of naïve T cells to Th17 which produces IL-17. *In vitro* studies show that high concentrations of retinoic acid inhibit differentiation of naïve CD4$^+$ T cells to Th17 by blocking the expression of retinoic-acid-receptor-related orphan receptor-γt (RORγt), a key transcription factor for Th17 cell differentiation. However, low concentrations of retinoic acid induce differentiation to Th17 cells.

TGF-β drives the development of iTreg cells in the periphery and this is enhanced by retinoic acid. iTregs are Treg cell populations that arise from precursor cells in the periphery as opposed to naturally occurring Tregs that originate from the thymus. iTregs can be also be found on the mucosal surfaces of the intestine. Retinoic acid has an impact on DCs from the small intestinal lamina propria or gut-associated lymphoid tissue (GALT) on enhancement of Treg differentiation. Tregs produce cytokines such as IL-10 which modulate inflammatory responses. Regulatory T cells play an important role in shutting down immune responses after successful elimination of pathogens, and can also prevent autoimmunity and immune tolerance to food antigens. Thus, retinoic acid maintains immunological tolerance by inducing differentiation of Treg cells and blocking Th17 cell differentiation at high concentrations. However, at low concentrations Th17 differentiation is promoted. Thus, the balance of Th17 and Tregs is influenced by the concentration of retinoic acid, with a dominance of Tregs at high retinoic acid concentrations.

4. **Effects on differentiation to Th1 cells**
   In response to infectious agents such as bacteria, retinoic acid enhances Th1 response and pro-inflammatory cytokine production.
   There is evidence that the balance between Th1 and Th2 cells is altered in vitamin A deficiency state; specifically, Th2 responses are decreased (Figure 2.5). However, the impact of retinoic acid on Th1 responses is less clear as studies have shown that Th1 responses could be unaffected, enhanced or inhibited.

5. **Increased homing of lymphocytes to the gut**
   Animal studies show that the production of retinoic acid by DCs plays an important role in intestinal immunity. Retinoic acid produced

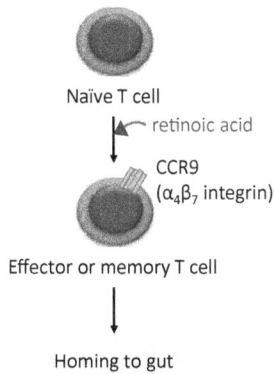

Figure 2.5   Proposed mechanism for homing of lymphocytes to the gut. Retinoic acid induces expression of CC-chemokine receptor 9 (CCR9) on T cells, and gut homing receptor α4β7 integrin (not shown) leads to homing of lymphocytes to the gut.

induces the homing of CD4$^+$, CD8$^+$ and B lymphocytes to the intestinal mucosae by inducing the expression of α4β7 integrin and the chemokine receptor CCR9 on both T and B cells (Figure 2.5). Depletion of vitamin A in the diet resulted in a decrease in the numbers of IgA-secreting B and T cells in the lamina propria. Thus, vitamin A metabolism by intestinal DCs and endothelial cells plays a key role in both T cell differentiation and homing of lymphocytes to maintain homeostasis in the gut.

6. **Induction of IgA class switching**

    Retinoic acid produced by mouse DCs in the gut has been shown to synergise with IL-5 and IL-6 to promote the secretion of IgA (Figure 2.6). The class switching to IgA was found to be dependent on mucosal DCs and also augmented by retinoic acid secreted by DCs. This is consistent with the observation that mice deficient in vitamin A lack T cells and IgA-secreting cells in the intestine. Vitamin A-deficient individuals thus have depressed IgA levels. *In vitro*, vitamin A can also suppress immunoglobulin class switching to IgG1 or IgE, which are isotypes associated with immune pathology and allergy. Thus, vitamin A plays a role in a balanced T cell response to infection and also maintenance of tolerance to allergens. However, high dosages of vitamin A may cause immunosuppression.

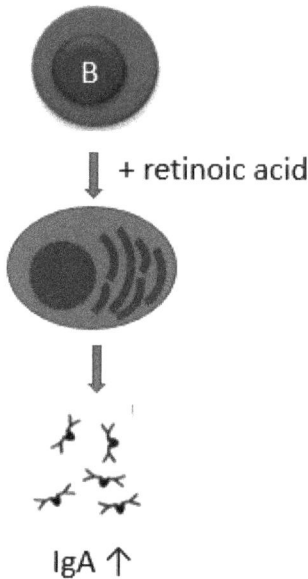

Figure 2.6   Retinoic acid induces IgA production from plasma cells.

2.2.6 *Vitamin A deficiency and susceptibility to infections*

Clinical signs of vitamin A deficiency include night blindness, mucosal damage, dry skin and hyperkeratosis. Vitamin A deficiency has been shown to result in increased pathogen replication and enhanced production of pro-inflammatory cytokines. In contrast, vitamin A supplementation has been shown to down-regulate the secretion of proinflammatory cytokines in response to infectious agents. Neutrophil functions related to chemotaxis, phagocytosis and intracellular killing are impaired in vitamin A deficiency states. NK cell numbers and ability to kill pathogens are also diminished.

Key points on functions of vitamin A:

- Maintains and restores the integrity and function of all mucosal surfaces
- Potentiates the antibody response especially IgA at the mucosa
- Regulates APCs
- Enhances MMP-9 expression
- Enhances T cell proliferation and differentiation to either Th2 or Tregs
- Induces IgA class switching

- Promotes the movement of T cells to the GALT
- Has the ability to maintain immune tolerance through interactions with DCs, GALTs, secretory IgA and gut microflora

Levels of serum vitamin A are altered by infection. For example, infection with *Schistosoma mansoni* depletes vitamin A. Deficiency in vitamin A and its metabolite retinoic acid in the diet has been associated with increased HIV pathogenesis, increased susceptibility to respiratory infections, diarrhoea and severe measles in children. This could be due to impairment of the activity of phagocytes, NK cells and Th2 cells. Thirty years ago, undernourished children were observed to have xerophthalamia and mortality rates were high. Vitamin A supplementation has been shown to reduce mortality to those diseases. Vitamin A deficiency can also impair response to vaccination, while supplementation increases antibody responses as was shown with tetanus toxoid and diphtheria vaccine in vitamin A-deficient children. Hence, vitamin A supplementation of children in developing countries has been found to prevent infectious disease and mortality as well as enhance vaccine responses.

It should be noted that the effects of vitamin A on immune cells are from *in vitro* or animal model studies. The clinical consequences of vitamin A deficiency are clear. Vitamin A supplementation is beneficial in deficient children as it increases vaccine responses, improves recovery from infectious diseases and decreases mortality. However, it is unclear if vitamin A is of therapeutic benefit in healthy humans as high doses may cause immunosuppression, and overdosing can lead to dermal dryness, loss of appetite and loss of hair. Nevertheless, there is potential for applications in vaccination or autoimmune conditions. Animal studies provide evidence that vitamin A metabolites are able to foster gut-homing T cells, which may improve strategies for mucosal vaccines or help decrease autoimmunity by potentiating the induction of regulatory T cells.

2.3 Vitamin B

2.3.1 *Introduction*

Vitamin Bs are water-soluble vitamins. Eight types of vitamin B have been described and their functions are as shown in the table below.

2.3.2   *Vitamin B and chronic diseases*

Micronutrients such as vitamins and minerals are constituents of a healthy diet and are important in the prevention of chronic diseases. Moderately elevated levels of homocysteine in the blood have been shown to increase the risk of cardiovascular diseases. Homocysteine exerts adverse effects on blood clotting, arterial vasodilation, and thickening of arterial walls. The breakdown of homocysteine to the amino acid methionine requires a vitamin B_6-dependent enzyme and a vitamin

Table 2.2   Functions of eight types of vitamin B.

Vitamin	Function	Effects of Deficiency
Vitamin B_1 (thiamine)	Thiamine is required for breakdown of carbohydrates to provide energy and is essential for nerve, muscle and heart functions.	Mild deficiency is associated with malaise, weight loss, poor sleep, irritability, and confusion. Severe deficiency is associated with beri-beri and heart failure.
Vitamin B_2 (riboflavin)	Vitamin B_2 is a component of cofactors in enzymes, such as flavin adenine dinucleotide, which are involved in the metabolism of macronutrients such as carbohydrates, proteins and fats to produce energy in the form of ATP.	Deficiency of vitamin B_2 causes ariboflavinosis, which is characterised by sores in the mouth. Deficiency of vitamin B_2 inhibits macrophage viability and weakens the ability of macrophages to bind to microorganisms.
Vitamin B_3 (niacin)	Vitamin B_3 plays a role in fuel metabolism and formation of red blood cells and skin. It can be found in two forms, nicotinic acid and nicotinamide. Nicotinamide is a component of nicotinamide adenine dinucleotide (NAD) and nicotinamide adenine dinucleotide phosphate (NADP), which are coenzymes involved in the metabolism of glucose and fat.	Severe deficiency of vitamin B_3, which is rare unless in under-developed countries, causes pellagra. Symptoms are related to the skin, digestive and nervous systems including dermatitis, diarrhoea, vomiting, mental confusion and dementia.

Table 2.2 (*Continued*)

Vitamin	Function	Effects of Deficiency
Vitamin B_5 (panthothenic acid)	Vitamin B_5 is the precursor of coenzyme A, which is involved in the synthesis of amino acids, fatty acids, cholesterol, phospholipids, steroid hormones, and neurotransmitters.	Deficiency of Vitamin B_5 is rare. It can result in fatigue, depression, prickling or burning of the skin, vomiting and stomach pains.
Vitamin B_6 (pyridoxine)	Vitamin B_6 acts as a coenzyme in the synthesis of proteins, neurotransmitters and nucleic acids. Vitamin B_6 is required for conversion of homocysteine to cysteine, production of antibodies, and in cell-mediated immunity.	Deficiency of vitamin B_6 impairs humoral and cell-mediated immunity. Failure of homocysteine conversion to cysteine results in an increased homocysteine level, which is associated with increased risk of heart disease.
Vitamin B_7 (biotin)	Biotin is a cofactor of coenzymes in the metabolism of lipids, proteins and carbohydrates.	Deficiency of vitamin B_7 is rare. Signs include dermatitis, lethargy and slight anaemia. Infants suffer from impaired growth and neurological disorders.
Vitamin B_9 (folate)	Vitamin B_9 is important in the production of red blood cells and cell growth. Vitamin B_9 is attached to transport proteins in the blood and haemoglobin of red blood cells. It acts as a co-enzyme which is involved in the metabolism of nucleic acids and amino acids. It is needed for cell division, especially during pregnancy and infancy.	Deficiency may be caused by inadequate dietary intake, malabsorption or increased requirements during pregnancy or haemolysis. When vitamin B_9 is deficient, red blood cell synthesis is inhibited leading to macrocytic anemia. Deficiency also results in elevated levels of homocysteine. High homocysteine levels are associated with increased risk of heart disease, and has also been found in patients with chronic immune-related diseases like rheumatoid arthritis or psoriasis. Deficiency in pregnant women can lead to birth defects. Folic acid supplementation is often recommended during pregnancy.

(*Continued*)

Table 2.2 (*Continued*)

Vitamin	Function	Effects of Deficiency
Vitamin B_{12} (cobalamin)	Vitamin B_{12} is also known as cobalamin as it contains cobalt. Vitamin B_{12} is a cofactor for only two enzymes, methionine synthase and L-methylmalonyl-coenzyme A mutase. Vitamin B_{12} is involved in carbohydrate, protein and lipid metabolism. It is essential for production of blood cells in bone marrow, and synthesis of myelin and maintenance myelin sheaths which protect the nerves.	Deficiency of vitamin B_{12} causes anaemia similar to that of vitamin B_9 deficiency. Deficiency results in the inability of methionine synthase to demethylate 5-methylhydrofolate and generate methylcobalamin. This causes secondary folate deficiency which results in megaloblastic anemia. Lack of methylcobalamin also causes elevated homocysteine levels which are associated with peripheral neuropathy, dementia and other cognitive deficits. Total number of lymphocytes and $CD8^+$ cells are decreased and NK cell activity are suppressed, which contribute to increased susceptibility to infections.

B_{12}-dependent enzyme with folate as a cofactor (Figure 2.7). Thus, these three vitamins play a key role in regulating the amount of homocysteine. Several studies have shown association between high concentrations of homocysteine and low concentrations of blood vitamin B_{12}, folate and vitamin B_6. Low concentrations of these vitamins have been linked with heart disease, stroke and other vascular outcomes. High homocysteine levels have also been found in patients with chronic immune-related diseases such as rheumatoid arthritis or psoriasis. It has been reported that treatment with folic acid is beneficial for patients with chronic inflammatory skin diseases such as psoriasis.

Key points on homocysteine:

- Actions of accumulated homocysteine:
  - promote blood clotting
  - increase arterial vasodilation
  - promote thickening of arterial walls
- Breakdown of homocysteine requires vitamin B_{12}-dependent enzyme, folate and vitamin B_6.
- Deficiency or low availability of folate, vitamin B_{12} or vitamin B_6 causes accumulation of homocysteine

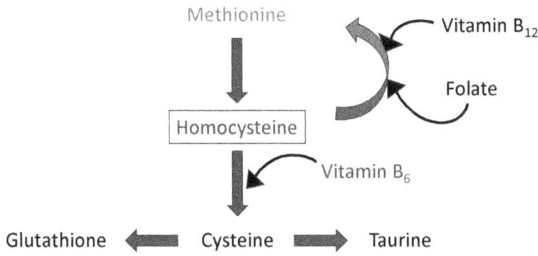

Figure 2.7 Folate, vitamins B_{12} and B_6 regulate blood homocysteine levels. Methionine is metabolised to homocysteine, which is then metabolised to cysteine in the presence of vitamin B_6. When vitamin B_{12} and folate are available, homocysteine is converted back to methionine. When vitamin B_{12}, folate and vitamin B_6 are deficient, homocysteine accumulates in the blood.

2.3.3 *Effects of vitamin B on the immune system*

A number of studies has shown that vitamin B has effects on the immune system. Deficiency of B_2 (riboflavin) inhibits macrophage viability and weakens the ability of macrophages to bind to microorganisms. Vitamin B_6 (pyridoxine) has been shown to be involved in the production of antibodies and in cell-mediated immunity, and is beneficial in the treatment of inflammatory infections of the respiratory tract. Vitamin B_{12} (cobalamin) has a role in regulation of the immune system. It has been shown to enhance *in vitro* concanavalin A-stimulation of T cell proliferation and pokeweed mitogen-induced immunoglobulin synthesis of B cells.

Key points on effects of vitamin B on the immune system:

- Effects of vitamin B_2 (riboflavin) deficiency on the immune system:
  — inhibition of macrophage viability
  — impaired adherence of macrophages to microorganisms
- Vitamins B_6 and B_{12} enhance production of antibodies and cell-mediated immunity

---

2.3.4 *Deficiency in vitamin B_{12}*

The most important function of vitamin B_{12} is DNA synthesis. Deficiency may be due to malabsorption, inadequate dietary intake or pernicious anaemia (antibodies to intrinsic factor). Vitamin B_{12} deficiency can occur in the elderly where absorption through the gut declines with age. It can cause the

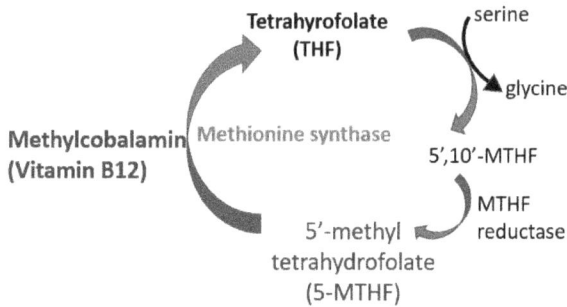

Figure 2.8   Metabolic pathway showing the role of methylcobalamin (vitamin B_{12}) and methionine synthase in the recycling of folate.

inability of methionine synthase to demethylate 5'-methyltetrahydrofolate (Figure 2.8), resulting in secondary folate deficiency and impairment of DNA and RNA synthesis. This leads to megaloblastic anaemia which can be corrected by folate supplementation. Deficiency in vitamin B_{12} also interferes with conversion of homocysteine to methionine. Increased homocysteine is associated with peripheral neuropathy. Vitamin B_{12}-deficient patients were found to have a decrease in the number of lymphocytes and $CD8^+$ cells, as well as suppressed NK cell activity. Deficiency in vitamin B_{12} can also lead to DNA damage in mononuclear leukocytes and this can be reduced by the administration of vitamin B_{12} injections. Animal model studies have shown that deficiency in vitamin B_{12} has a suppressive role in protection against viral and bacterial infections. This observation is consistent with inhibition of NK cell activity and reduced number of lymphocytes in the blood of vitamin B_{12}-deficient patients.

Key points on deficiency in vitamin B_{12}:

- Causes secondary folate deficiency leading to megaloblastic anaemia
- Causes peripheral neuropathy
- Decreases lymphocyte numbers
- Decreases NK cell activity
- Increases susceptibility to viral and bacterial infections

Humans obtain vitamin B mostly from diet and are unable to synthesise it unlike commensal bacteria. The by-products from commensal bacterial-synthesised vitamin B can activate a special type of T cells known as MAIT cells which are abundant on mucosal surfaces. Thus, commensal bacterial-synthesised vitamin B can possibly play a role in protective immunity against infections.

2.4 Vitamin C

2.4.1 *Introduction*

Vitamin C, also known as ascorbic acid, is a water-soluble vitamin. It has to be obtained from the diet as it cannot be synthesised by the body. Inadequate intake of vitamin C causes scurvy, which is characterised by bleeding gums, swollen and painful legs, bruising, skin haemorrhages, fatigue and apathy. In the 16th to 18th century, scurvy was associated with extended sea travels. Dietary sources of vitamin C are fruits and vegetables such as citrus fruits, tomatoes, strawberries, kiwi fruit, broccoli, red and green peppers and cantaloupe.

Key points on functions of vitamin C:

- Acts as an antioxidant preventing cell membrane and DNA damage
- Maintains skin integrity
- Involved in collagen synthesis
- Helps in iron absorption
- Stimulates the production of Type I interferon which has antiviral action
- Promotes phagocyte chemotaxis
- Increases T lymphocyte proliferation

2.4.2 *Functions of vitamin C*

Vitamin C has immune-stimulating effect and neutralises free radicals such as reactive oxygen species. Since vitamin C is water-soluble, it is an important antioxidant in the extracellular fluid. Upon phagocytosis of pathogen, oxidative burst occurs resulting in generation of reactive oxygen species such as superoxide ($O_2^{\bullet}$) and eventual killing of engulfed pathogen. However, the reactive oxygen species are harmful to host cells as they have pro-oxidant effects that cause cell membrane and DNA damage. High concentrations of reactive oxygen species can cause oxidative stress which leads to degenerative disorders such as aging, stroke, cardiovascular

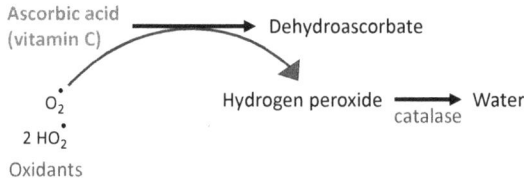

Figure 2.9   Antioxidant function of ascorbic acid (vitamin C).

disease, cancer, cataract and macular degeneration. Vitamin C prevents oxidative damage caused by reactive oxygen species in immune cells (Figure 2.9). Vitamin C also provides antioxidant protection for plasma lipids and lipid membranes. Thus, vitamin C plays an important role as an antioxidant that protects various cells against oxidative damage.

Normal vitamin C intake is vital to skin integrity which is important for immune function. Vitamin C acts as a cofactor for enzymes known as hydroxylases for collagen synthesis. Collagen is the most abundant protein in the body and has an important role in wound healing. Vitamin C also helps in iron absorption in the intestine, and can promote the activity of leukocytes. Experimental models have shown that vitamin C promotes chemotaxis of phagocytes and stimulates interferon production, which interferes with viral replication.

The concentration of vitamin C in phagocytes and lymphocytes is higher than in plasma but it is reduced in infections as it is rapidly utilised.

2.4.3 *Vitamin C supplementation*

Vitamin C supplementation at doses of 1–5 g daily for 3 days has been shown to increase proliferation of human T lymphocytes. Megadoses of vitamin C are not recommended. There is weak evidence to support the use of vitamin C prophylactically. However, vitamin C may be beneficial for treating community-acquired pneumonia if the patient has low vitamin C levels. Vitamin C has also been shown to reduce the duration and severity of infections of the upper respiratory tract in some studies. Evidence also indicates that regular consumption of vitamin C can combat colds for workers who do laborious physical work and/or are exposed to cold environments for long periods of time. Supplementation with 200 mg of vitamin C per day was found to ease the symptoms of acute respiratory infections in the elderly. Finally, increased neutrophil chemotaxis and improved bactericidal activity has been shown in Chediak-Higashi syndrome patients treated with vitamin C. Thus, vitamin C is important for its antioxidant effects and enhances immunity in immunocompromised individuals such as the elderly, and those suffering from infectious diseases and genetic immunodeficiency.

Key points on evidence to support use of vitamin C:

- Reduced duration and severity of infections of the upper respiratory tract
- Combat colds for workers who do hard physical work over short periods of time and/or are exposed to low temperatures for lengthy periods
- Alleviate the symptoms of acute respiratory infections in the elderly
- Increased neutrophil chemotaxis and improved bactericidal activity have been shown in those afflicted with Chediak-Higashi syndrome

---

2.5 Vitamin D

2.5.1 *Introduction*

Vitamin D was first identified and characterised in 1923 by Goldblatt and Soames. Vitamin D exists in three forms, namely, secosterols, ergocalciferol (vitamin D_2), and cholecalciferol (vitamin D_3). The most physiologically relevant vitamin D is vitamin D_3, which can be acquired in the diet, vitamin supplements or synthesised in the skin from 7-dehydrocholesterol under sunlight. It is hydroxylated in the liver and cells of the immune system to 25-hydroxyvitamin D, then hydroxylated to 1,25-dihydroxyvitamin D in the kidneys. 1,25-dihydroxyvitamin D is the physiologically most active metabolite of vitamin D. It is transported into the blood where it has multiple systemic effects.

For instance, it acts on immune cells including DCs, macrophages, T and B cells in an autocrine or paracrine manner by binding to the vitamin D receptor, leading to transcriptional activation of genes. It is eventually catabolised to calcitroic acid, which is excreted in the bile.

2.5.2 *Dietary sources of vitamin D*

Lean meats, poultry, and oily fish such as mackerel, salmon, sardines, beans, cod liver oil and eggs are rich in vitamin D. Other sources include fortified breakfast cereals, cheese and powdered milk.

2.5.3 *Vitamin D deficiency*

Severe vitamin D deficiency in infants and children results in the failure of bone to mineralise, leading to bone deformities (a condition referred to as rickets). In adults, progressive loss of bone mineral results in softening of bones (osteomalacia), bone pain, and increased risk of osteoporosis. Vitamin D deficiency can also cause muscle weakness and pain in children and adults. The reduction of intestinal calcium absorption in vitamin D deficiency results in increased parathyroid hormone secretion by the parathyroid gland. This may result in increased bone resorption, leading to possible increased risk of osteoporotic fracture.

Low serum vitamin D levels have been reported for certain cancers such as breast, prostate and colorectal cancer. However, other studies do not support these findings so it is still unclear whether vitamin D supplementation is beneficial for reducing cancer risk.

2.5.4 *Effects of vitamin D on immune cells*

1,25-dihydroxyvitamin D_3 ($1,25(OH)_2VD_3$) enhances innate responses to some bacteria by inducing monocyte proliferation and the production of cathelicidin (an antibacterial peptide) and IL-1 by macrophages. It promotes phagocytosis, superoxide production and killing of bacteria (Figure 2.10).

$1,25(OH)_2VD_3$ decreases DC maturation, and inhibits antigen presentation by inhibiting upregulation of MHC class II and costimulatory molecule expression. In addition, it decreases IL-12 (a cytokine that drives differentiation to Th1 cells) production by DCs while inducing the production of IL-10.

$1,25(OH)_2VD_3$ inhibits proliferation of T cells, decreases IL-2 and interferon-γ (IFN-γ) production from T cells, and attenuates $CD8^+$ cytotoxic activity. *In vitro* studies show that $1,25(OH)_2VD_3$ inhibits polarisation of naïve T cells to Th1 cells, induces polarisation to Th2 cells and can also promote the development of Treg cells (Figure 2.11). $1,25(OH)_2VD_3$ also inhibits Th17 differentiation and cytokine production by Th17 cells. In addition, $1,25(OH)_2VD_3$ inhibits B cell proliferation, plasma cell differentiation and immunoglobulin production. Thus, $1,25(OH)_2VD_3$ appears to have inhibitory effects on T and B lymphocytes, and stimulatory effects on innate cells such as macrophages.

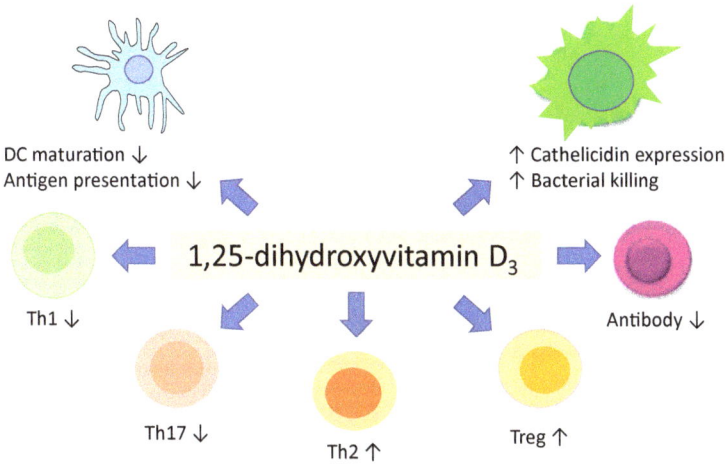

Figure 2.10 Effects of $1,25(OH)_2VD_3$ on immune cells.

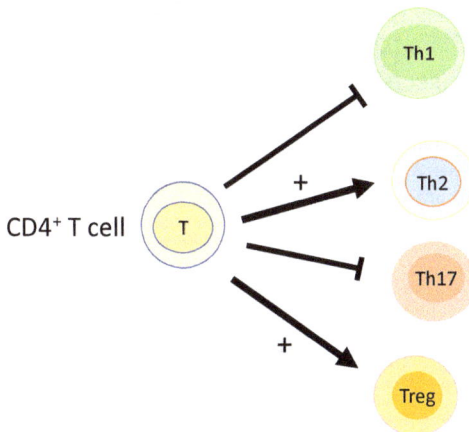

Figure 2.11 Effect of $1,25(OH)_2VD_3$ on differentiation of naïve $CD4^+$ T cells.

$1,25(OH)_2VD_3$ induces homing of lymphocytes to the skin by inducing the expression of CCL27 on keratinocytes of the skin. CCL27 is a chemoattractant for CCR10 expressing cells. $1,25(OH)_2D_3$ induces the expression of CCR10 on T cells. The migration of $CCR10^+$ T cells towards the skin is significantly induced in the presence of $1,25(OH)_2D_3$ and IL-12 (Figure 2.12).

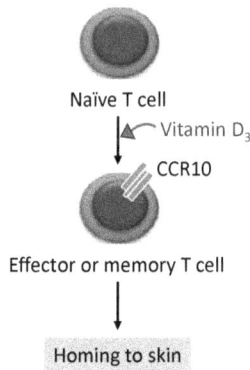

Figure 2.12   Homing of T cells to the skin is induced by $1,25(OH)_2VD_3$.

2.5.5 *Vitamin D supplementation in chronic diseases*

Epidemiologic studies have associated vitamin D deficiency with increased risk of chronic inflammatory diseases and auto-immune diseases. These diseases include cancers such as colon, prostate, and breast cancer, cardiovascular diseases, diabetes, multiple sclerosis and rheumatoid arthritis.

Key points on benefits of vitamin D supplementation:

- Reduces susceptibility to infections, e.g. influenza, tuberculosis
- Lowers risk of autoimmune diseases, e.g. Type I diabetes

Epidemiological, experimental, and interventional evidence have provided support for the health-promoting roles of vitamin D. Dietary supplementation and controlled sun exposure can improve long-term health and resistance to infections and chronic diseases. Individuals with low vitamin D status have a higher risk of respiratory tract viral infection. A study in Japan showed that vitamin D supplementation of school children for 4 months during winter decreased the risk of influenza by 40%. Vitamin D supplementation has also shown benefits in patients with tuberculosis and chronic obstructive pulmonary disease, and is associated with lower risk of developing autoimmune disorders. These studies suggest that vitamin D can reduce susceptibility to infections.

Key points on effects of 1,25(OH)$_2$VD$_3$ on immune function:

- Enhances innate immunity: induces monocyte proliferation and production of cathelicidin (an antibacterial peptide) and IL-1γ by macrophages
- Promotes phagocytosis, superoxide production and killing of bacteria
- Decreases DC maturation and inhibits upregulation of MHC class II and costimulatory molecule expression
- Decreases production of IL-12 (a Th1-promoting cytokine) by DCs while inducing the production of IL-10
- Inhibits T cell proliferation, decreases production of Th1 cytokines (IFN-γ) from T cells and attenuates the activity of cytotoxic T cells
- Inhibits polarisation to Th1 cells
- Induces polarisation to Th2
- Inhibits Th17 differentiation and cytokine production by Th17 cells
- Promotes the development of Treg cells
- Inhibits B cell proliferation, plasma cell differentiation and immunoglobulin production
- Inhibits the development of autoimmunity

However, a recent study showed no reduction in incidence or severity of upper respiratory tract infections in healthy adults. There is also literature that shows that vitamin D and its analogues have immunosuppressive effects. Thus, whether vitamin D supplementation is beneficial depends on the immunological status of the individual (i.e. healthy or suffer from infectious diseases and autoimmune diseases).

In recent years, the role of vitamin D in the pathogenesis and prevention of type I and II diabetes has been of great interest. Autoimmune destruction of insulin-producing β-cells in the pancreas is a characteristic feature of type I diabetes. Animal studies show that 1,25(OH)$_2$D$_3$ had significantly reduced the incidence of diabetes due to prevention of β-cell damage caused by pro-inflammatory cytokines (IL-1β and IFN-γ). Synthesis of these cytokines are inhibited by 1,25(OH)$_2$D$_3$. Children who received vitamin D supplementation were at reduced risk of developing childhood-onset diabetes type I. This could be due to the immune suppressive role of 1,25(OH)$_2$D$_3$. *In vitro* studies show that 1,25(OH)$_2$D$_3$ can downregulate antigen presentation and expression of co-stimulatory molecules, inhibit production of pro-inflammatory cytokines and promote the induction of regulatory T-lymphocytes. In type II diabetes, vitamin D has been shown to improve insulin secretion and glucose tolerance. However, studies on vitamin D supplementation on prevention of type II diabetes are still inconclusive. Further clinical trials are required to support a role for vitamin D supplementation in the prevention of type II diabetes.

2.6 Vitamin E

2.6.1 *Introduction*

Vitamin E is a fat-soluble vitamin which consists of four forms of tocopherol and four forms of tocotrienols. α-tocopherol is the most biologically active form of vitamin E and can be found most abundantly in sunflower, wheat germ oil and safflower oils. γ-tocopherol can be found in margarine, corn oil, soybean oil, and dressings.

Vitamin E is important to immune function as an antioxidant and as a cell membrane component. Vitamin E protects polyunsaturated fatty acids and other fatty components of the cell and cell membranes from being oxidised (Figure 2.13). The prevention of low-density lipoprotein oxidation lowers the risk of heart disease. It also protects leucocytes and other components of the immune system, and improves the absorption of vitamin A if dietary vitamin A level is low.

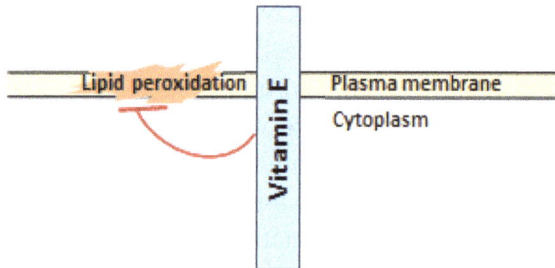

Figure 2.13 Antioxidant action of Vitamin E. Oxidative degradation of membrane lipids (lipid peroxidation) by reactive oxygen species results in cell damage. Vitamin E is a major lipid-soluble antioxidant and it can prevent lipid peroxidation.

2.6.2 *Effects of vitamin E on cells of the immune system*

Experimental findings indicate that vitamin E supplementation can stimulate the immune system directly in the following manner:

(1) Vitamin E increases lymphocyte proliferation in response to antigen and production of IL-2, increased NK cell activity and phagocytosis. Vitamin E also increases antibody production.
(2) Vitamin E can decrease monocyte adhesion to the endothelium by preventing upregulation of the adhesion molecules on the endothelium.
(3) Vitamin E decreases the formation of pro-inflammatory cytokines in macrophages (scavenger cells) and monocytes. Decreased inflammation by vitamin E is due to blockade in the release of pro-inflammatory cytokines, including TNF, IL-1, IL-8 and IL-6 by monocytes and macrophages.
(4) Vitamin E can act on mast cells. Mast cells are activated by oxidised lipoproteins that occur under the influence of pro-inflammatory cytokines. These cells then release mediators that promote allergic inflammatory processes. Excessive production and activation of mast cells prevented by the antioxidant properties of vitamin E can limit the oxidation of lipoproteins. In this way, vitamin E can dampen overreactions of the immune system and allergic reactions.

2.6.3 *Effects of vitamin E deficiency*

In humans, severe vitamin E deficiency is associated with weakened T lymphocyte function. Increased intake of vitamin E can have protective antioxidant and immune-stimulating effects as well as decrease inflammation in degenerative diseases such as atherosclerosis. In healthy older adults with high plasma vitamin E levels, there is enhanced cell-mediated immunity responses and reduced risk of infection. The benefits of vitamin E is as evidenced from vitamin E deficiency, which is associated with increased susceptibility to infections. There is evidence to support the benefits of vitamin E supplementation in the elderly where there is enhanced Th1 cell-mediated immunity, decreased risk of upper respiratory infections and improved vaccination responses. However, there are other studies which did not see an effect on the incidence, duration or severity of respiratory infections in the elderly.

In conclusion, there is experimental evidence to show that vitamin E exerts protective antioxidant and immune-stimulating effects, as well as decreases inflammation in allergy and degenerative diseases such as atherosclerosis. It should be noted that high doses of vitamin E may depress phagocytosis and intracellular killing of bacteria.

Key points on effects of vitamin E:

- Antioxidant effects — limits the oxidation of lipoproteins and prevents the release of reactive oxygen species by phagocytes
- Decreases inflammation by
  - blocking release of pro-inflammatory cytokines such as TNF, IL-1, IL-6 and IL-8 from monocytes and macrophages
  - preventing upregulation of adhesion molecules on endothelium, resulting in decreased monocyte adhesion to endothelium
  - decreasing IFN-γ production by T lymphocytes, resulting in decreased inflammation and tissue damage
- Stimulates immunity by
  - increased T cell proliferation
  - increased IL-2 production
  - increased antibody production
  - increased NK activity
  - increased phagocytosis by macrophages

Summary of Chapter 2

This chapter introduces the different forms of vitamins A, B, C, D and E and provides examples of dietary sources. The benefits of these vitamins in humans are supported by the clinical evidence from specific vitamin-deficient patients. The effects of these vitamins on immune cells are then described. Vitamin A is important in maintaining epithelial and mucosal barrier functions, phagocytic and NK cell activity, enhancing Th2 responses, potentiating the antibody response especially IgA at the mucosa, regulating APCs, inducing Tregs in combination with TGF-β and promoting homing of T cells to the GALTs. Eight types of vitamin

B are described and their functions are summarised in Table 2.2. Vitamins C and E are well-known for their antioxidant effects. In addition, vitamin E has immune-stimulatory effects and decreases inflammatory effects. The active form of vitamin D can induce macrophages to produce antibacterial peptides, promote phagocytosis but inhibit T cell proliferation, produce Th1 cytokines and antibodies, and inhibit autoimmunity. Thus, these vitamins regulate the immune response and dietary intake of vitamins are essential for immune function.

Chapter 3

Minerals and immune function

Learning objectives

After studying this chapter, you should be able to explain:

1. Basics of minerals, dietary sources and antioxidant functions
2. Iron metabolism and how copper is involved
3. Functions of magnesium and iron
4. Effects of mineral deficiency on immune function
5. Roles of copper, zinc and manganese
6. How zinc influences copper overload
7. Importance of selenium in susceptibility to infection

3.1 Introduction

Minerals are inorganic substances needed by the body in small quantities. There are two types of minerals, namely, macrominerals and microminerals. Examples of macrominerals are calcium, phosphorus, potassium, sulphur, sodium, chloride and magnesium. Microminerals (trace minerals) include iron, zinc, copper, manganese, iodine, selenium, chromium, fluoride, and molybdenum. Zinc, selenium, copper, and iron play essential roles in the immune responses.

Calcium is important for the formation and metabolism of bone. Over 99% of total body calcium is in bones and teeth. Calcium present in blood, extracellular fluids, muscles and other tissues play important physiological roles such as vascular contraction and vasodilation, secretion of hormones, nerve transmission and muscle function. Vitamin D aids in absorption of calcium. In vitamin D deficiency states, reduced calcium absorption affects bone metabolism. Calcium metabolism is regulated by the parathyroid hormone-vitamin D system. 75% of dietary calcium is obtained from dairy products.

Examples of dietary sources of minerals are as follows:

- Calcium: milk, yogurt, cheeses, legumes (such as dried peas and beans), dark green leafy vegetables
- Chromium: red meats, whole grains, spices
- Copper: liver, green vegetables, soybeans, tofu, potatoes
- Fluoride: fluoridated water
- Iodine: seafood, milk, cheeses, cereals, iodised salt

The body maintains tight control over mineral balance. Proper mineral ratios are required for basic good health. Copper/zinc ratio of 1:1 and calcium/magnesium ratio of 2:1 are needed. The gastrointestinal tract regulates absorption from food based on the needs of the body. Minerals in gastric juices are either excreted in the faeces or reabsorbed through the large intestine. Excess minerals are secreted by the kidneys and reabsorbed when required by the body.

3.2 Functions of minerals: Overview

The functions of minerals are to: (1) serve as cofactors for many enzymes, (2) maintain fluid and electrolyte balance, (3) renew and support the function of red blood cells, (4) build healthy bones, and (5) maintain a healthy immune system.

3.2.1 *Minerals as cofactors for enzymes*

A cofactor is defined as a substance that helps catalyse a reaction. Minerals serve as cofactors in the following systems: (1) metabolism of carbohydrates and fats for energy production, (2) antioxidant systems, (3) muscle contraction, and (4) nerve transmission.

3.2.2 *Minerals as cofactors in antioxidant systems*

The term "oxidative stress" has been used to define a state in which there is excess reactive oxygen species (ROS) and reactive nitrogen species (RNS), either due to excess production or insufficient removal. When pathogens such as bacteria are engulfed by phagocytes, respiratory burst generates ROS such as hydrogen peroxide that kill the engulfed pathogen (Figure 3.1). The oxidative bacterial damage occurring in the macrophages contributes to the defense against invading pathogens. However, excessive amounts of ROS and RNS can cause damage to proteins, lipids and DNA. Oxidative damage also occurs as a result of metabolic processes and environmental sources such as air pollution and cigarette smoke.

Key points on minerals as cofactors in antioxidant systems:

- Antioxidants are defined as substances that are able to neutralise reactive molecules and reduce oxidative damage
- Oxidative damage is a result of metabolic processes and environmental sources such as air pollution and cigarette smoke
- Vitamin C, vitamin E, beta-carotene, vitamin A are known antioxidants
- Minerals that act as cofactors include selenium, iron, zinc, copper and manganese
- Copper, zinc and manganese are part of the enzyme superoxide dismutase
- Iron is a cofactor for the catalase antioxidant enzyme
- Selenium is required for glutathione peroxidase activity

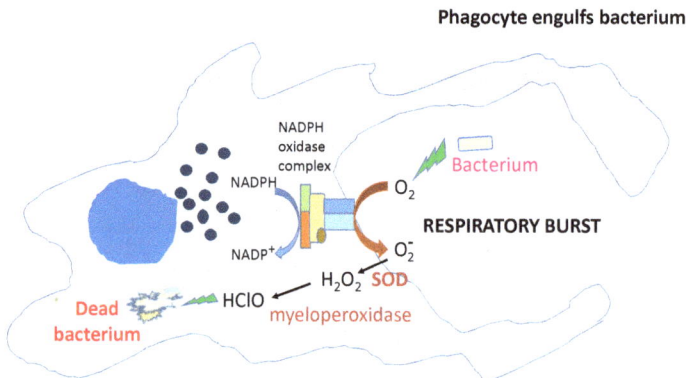

Figure 3.1   Generation of reactive oxygen species in respiratory burst during phagocytosis.

$$\cdot O_2^- \xrightarrow[\text{Mn-SOD}]{\substack{\text{Cu/Zn-SOD} \\ \text{(cytosol)}}} H_2O_2$$

Cu/Zn-SOD (cytosol)
Mn-SOD (mitochondria)
catalase | Glutathione Peroxidase (mitochondria)

$$H_2O + O_2$$

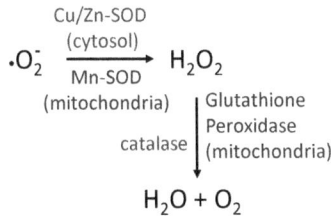

Figure 3.2 Antioxidant function of superoxide dismutase, catalase and glutathione peroxidase.

Vitamin C, vitamin E, β-carotene, and vitamin A are known antioxidants. Minerals act as cofactors for antioxidant enzymes. These vitamins and minerals remove potentially damaging ROS produced during oxidative stress to harmless products by coupling their reduction with the oxidation of glutathione. Thus, damage to lipids, proteins, and DNA is reduced.

The primary antioxidant defence system involving minerals as cofactors are the enzymes superoxide dismutase, catalase and glutathione peroxidase (Figure 3.2).

Superoxide Dismutase (SOD):

SODs are a group of metalloenzymes which catalyze the conversion of superoxide anion to hydrogen peroxide.

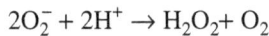

$$2O_2^- + 2H^+ \rightarrow H_2O_2 + O_2$$

Copper/zinc and manganese are cofactors for SODs in the cytosol and mitochondria, respectively.

Catalase

High amounts of hydrogen peroxide formed by SOD are scavenged by catalase. It catalyses the dismutation of hydrogen peroxide into water and molecular oxygen.

$$H_2O_2 \rightarrow O_2 + 2H_2O$$

Iron is a cofactor for the catalase antioxidant enzyme.

Glutathione Peroxidase

Glutathione peroxidase catalyses the reduction of low amounts of H_2O_2 using glutathione (GSH) as a substrate. They can also reduce other peroxides (e.g. lipid peroxides in cell membranes) to alcohols.

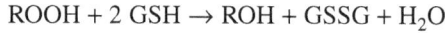

$$ROOH + 2\ GSH \rightarrow ROH + GSSG + H_2O$$

Selenium is required for glutathione peroxidase activity.

There is also an important intracellular non-enzymatic defence system involving GSH to protect cellular constituents against ROS and for maintaining the redox state. GSH is the most abundant intracellular thiol-based antioxidant and is an important antioxidant in our body.

The ratio of reduced-to-oxidised glutathione (GSH/GSSG) in normal cells is high (>10:1). Glutathione reductase reduces GSSG to generate GSH in the following reaction:

$$GSSG + NADPH + H^+ \rightarrow 2\ GSH + NADP^+$$

GSH is made up of three amino acids, namely, glycine, cysteine and glutamic acid. Supplementation with GSH precursors (L-cysteine and L-glycine) was found to lower oxidative stress and prevented or delayed onset of cardiovascular disease, and loss of hearing, eyesight and brain function in the elderly.

3.3 Magnesium

3.3.1 *Introduction*

Magnesium is an essential mineral and is a cofactor for more than 300 enzymes such as Na^+/K^+ ATPase, hexokinase, creatine kinase, protein kinase, and cyclases. Magnesium plays an important role in key physiological processes including nucleic acid and protein synthesis, energy production, calcium and potassium transport, and structural functions in cell membranes and bones. It is also essential for the regulation of blood pressure, cardiac excitability, muscular contraction, insulin metabolism, nerve transmission and neuromuscular conduction. Low levels of magnesium have been associated with a number of chronic diseases, such as type II diabetes mellitus, hypertension, stroke, cardiovascular disease, migraine headaches, and Alzheimer's disease.

Food sources for magnesium include whole grains, almonds, brazil nuts, green leafy vegetables, fish, meats and milk.

Individuals that are obese or have chronic diseases (such as heart disease and type II diabetes mellitus) are commonly found to be magnesium-deficient. Epidemiological studies have also linked higher intakes of magnesium with lower incidence of respiratory problems. In addition, severe magnesium deficiency can obstruct calcium and vitamin D homeostasis. Thus, magnesium deficiency should be a concern for health and well-being. Individuals with gastrointestinal diseases such as Crohn's disease or renal disorders, those suffering from chronic alcoholism, and older people are at higher risk of magnesium deficiency. In older people, the amount of magnesium declines as there is less absorption in the intestines, and less hydrochloric acid produced in the stomach affects the breakdown and absorption of vitamins and minerals from the foods consumed.

3.3.2 *Immune functions of magnesium*

The functions of magnesium in immune function are mainly derived from animal studies. Magnesium deficiency in animals is associated with impairment of cell-mediated and humoral immunity. Immunoglobulin IgG level is reduced and IgE is increased. Magnesium is required for the adhesion-recognition phase of cytotoxic T lymphocyte action, T and B cell adherence, complement activation and macrophage response to cytokines. Magnesium deficiency is accompanied by activation of macrophages, neutrophils and endothelial cells. Animal studies also show that magnesium deficiency induces inflammatory stress as measured by increased blood levels of pro-inflammatory cytokines. In humans, serum C-reactive protein concentrations (a widely used indicator of inflammation) were elevated in magnesium deficiency.

3.4 Iron

3.4.1 *Introduction*

Iron exists in two main forms: ferric (Fe^{3+}) and ferrous (Fe^{2+}). It is required by all living cells. Iron is bound to metalloproteins, enzymes, storage and transport proteins (e.g. transferrin) in the body. Foods rich in

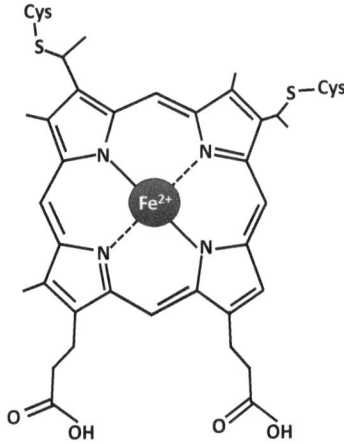

Figure 3.3   Structure of haem. Oxygen reversibly binds to the iron held in the centre of the porphyrin ring of the haemoglobin molecule.

iron include eggs, meat, liver and dried fruits. Iron in meat and liver is in the ferrous (Fe^{2+}) state. This form of iron is absorbed more readily than inorganic iron found in vegetables, grains and supplements which is in the ferric (Fe^{3+}) state. The haem iron is part of haemoglobin in blood (Figure 3.3) and myoglobin in animal flesh.

3.4.2 *Iron metabolism*

Iron absorbed from the intestine can be stored in intestinal epithelium and also in macrophages as ferritin or transported in plasma as transferrin. Plasma transferrin is a source of iron for haemoglobin synthesis. Iron can also be obtained by recycling of senescent erythrocytes by macrophages in the spleen and liver.

Key points on iron:

Iron is essential for

- binding and transport of oxygen
- generation of ATP in the mitochondria
- DNA synthesis and hence lymphocyte proliferation (rate limiting enzyme is ribonucleotide reductase, an iron metalloenzyme)
- enzymes that generate peroxide such as myeloperoxidase, which catalyses the production of toxic hydroxyl radicals to kill bacteria engulfed by neutrophils

3.4.3 *Effects of iron deficiency on the immune system*

Iron deficiency occurs in developing countries due to a scarcity of foods containing bioavailable iron. In developed countries, iron deficiency can occur due to inadequate dietary intake and loss of iron through menstruation. Iron impairs immune responses to invading pathogens. Anaemic children supplemented with iron had fewer infections. However, since iron is required for the survival and growth of several intracellular pathogens, including *Mycobacterium bovis*, one way to fight these pathogens is to restrict their access to iron. Animal studies have shown that survival of intracellular pathogens that cause tuberculosis or malaria were enhanced by iron therapy. This is consistent with the effects of high dose iron supplementation, whereby increased risk of clinical malaria and other infections such as pneumonia in children were reported. Thus, it is important to maintain an appropriate concentration of iron to generate an effective immune response and avoid sustained consumption of large doses of iron as it may induce tissue damage mediated by ROS.

Key points on effects of iron overload on the immune system:

- Reduced phagocytosis by neutrophils
- Phagocytosis by macrophages not affected but there is reduction in TNF and IL1 secretion
- Reduced T lymphocytes, reduced CD4+:CD8+
- Reduced proliferation of T lymphocytes
- No effect on humoral immunity and B cell function

3.5 Zinc

3.5.1 *Dietary sources of zinc*

The main food sources of zinc are spinach, garlic, pumpkin seeds, sunflower seeds, wheat germ, whole grains, meat, seafood and brewer's yeast. In the body, zinc together with vitamin B_6 and magnesium are used to synthesise γ-linolenic acid which is metabolised to prostaglandins.

3.5.2 *Absorption, transportation and excretion of zinc*

Like iron, zinc shows a shift from circulation into intracellular compartment during an infection. Zinc is not stored but is recycled. Zinc can be excreted in faeces, so it is less toxic than iron.

$$\cdot O_2^- \xrightarrow[\substack{\text{Mn-SOD} \\ \text{(mitochondria)}}]{\text{Cu/Zn-SOD}} H_2O_2$$

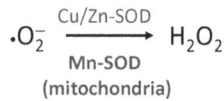

Figure 3.4   Zinc as a cofactor in SOD. SOD scavenges superoxide radical, a primary ROS to prevent oxidative damage.

3.5.3 *Functions of zinc: An overview*

There are at least 300 different enzymes which require zinc. These enzymes are involved in DNA synthesis, immune function, growth and development, haem synthesis and antioxidant systems. For instance, zinc is needed for insulin storage in the pancreas. An example of an enzyme that protects against oxidative damage in the body is the copper-zinc SOD (Figure 3.4).

3.5.4 *Effects of zinc on the immune system*

The essential role of zinc in immune function is evidenced from patients with zinc deficiency where both innate and adaptive immune response are impaired. Deficiency in zinc also causes low appetite and impaired growth. Individuals with deficiency in zinc suffer from diarrhoea, increased susceptibility to infections and skin lesions.

Key points on effects of zinc on the immune system:

- Maintains integrity of mucosal membranes and skin
- Inhibits virus replication
- Essential cofactor for thymulin which controls cytokine release and induces proliferation of immune cells
- Supports Th1 response
- Essential cofactor for cytosolic antioxidant enzyme SOD, which scavenges superoxide radical, a primary ROS, to prevent oxidative damage

---

Adequate zinc intake is important in innate immunity as it maintains skin and mucosal membrane integrity. Impairment of mucosal immunity allows entry of gut pathogens, leading to infections and diarrhoea.

Zinc-immunodeficient individuals suffer from lymphopenia and thymic atrophy, as well as have altered T-lymphocyte subsets and cytokine response profiles. This can be due to deficiency of zinc which is an essential cofactor for thymulin, which controls cytokine release and induces maturation and activity of immune cells. Zinc also supports Th1 response, T and B cell proliferation and antibody production.

3.5.5 Zinc as an antioxidant

Zinc is involved in defence against oxidative stress. It is an essential cofactor for a cytosolic antioxidant enzyme SOD that scavenges superoxide radicals, an ROS, to prevent oxidative damage to cells. Zinc synergises with vitamin C to protect immune cells against oxidative damage by ROS.

3.5.6 Zinc influences vitamin A metabolism

Zinc plays a regulatory role in transport of vitamin A. The retinol-binding protein which transports vitamin A from the blood to the liver requires zinc. Zinc deficiency can reduce the synthesis of retinol-binding protein in liver. Zinc is also involved in the oxidative conversion of retinol to retinaldehyde (retinal). This enzymatic conversion requires a zinc-dependent retinol dehydrogenase enzyme and is an important step needed for visual processes. Night blindness, which is a hallmark sign for vitamin A deficiency, is also seen with zinc deficiency.

3.5.7 Effects of zinc deficiency

A severe deficiency of zinc can have a drastic reduction in the activity of immune cells and antibody production. In the elderly, even a mild deficiency in zinc can increase their susceptibility to virus infections and allergic diseases. Several studies have shown that supplementation with zinc can reduce this susceptibility and strengthen the immune system. Zinc supplementation enhances phagocytosis by phagocytes, generation of oxidative burst, natural killer cell activity, antibody production and the numbers of cytotoxic $CD8^+$ T cells.

Key points on effects of zinc deficiency on immune cells:

- Impaired macrophage functions (phagocytosis and intracellular killing are impaired)
- Phagocytosis, generation of oxidative burst, and chemotaxis are reduced in neutrophils
- Natural killer cell activity is reduced
- Normal functions of T-cells are impaired (diminished proliferation, decreased ratio of CD4/CD8)
- B cell proliferation and antibody responses are suppressed

The supplementation of zinc in deficient individuals resulted in reversal of impaired immune functions including cytokine production, and reduced the incidence of diarrhoea and pneumonia in both adults and children.

However, overdose of zinc has negative effects on the immune status and also can cause copper deficiency.

3.6 Copper

3.6.1 *Sources of dietary copper*

Liver, seafood, shellfish, dark chocolate, dark leafy vegetables, soya, nuts, and mushrooms are sources of dietary copper. Copper deficiency is rare.

3.6.2 *Absorption, transportation and excretion of copper*

A protein produced by the liver known as ceruloplasmin binds copper for transport in blood. Little copper is stored and mostly excreted in faeces. Copper binds to metallothionein in the intestine and is excreted. Copper balance is controlled by copper absorption and influenced by iron and zinc statuses.

3.6.3 *Functions of copper*

Copper was first identified as an essential micromineral in the 1960s in a study on Peruvian children presented with iron-refractory anaemia. Copper is needed to help haemoglobin carry oxygen throughout the body. Other clinical signs were neutropenia (a diminished number of neutrophils) and bone abnormalities. The most profound effect of reduced number of neutrophils is increased susceptibility to infections. A genetic disorder for severe copper deficiency is Menkes disease. These infants suffer from frequent and severe infections.

Copper is an essential cofactor of a number of antioxidant enzymes such as cytosolic SOD, catalase and glutathione peroxidase in defence against ROS.

Copper deficiency also has a pronounced effect on the adaptive immune system. Copper deficiency may cause a decreased IL-2 response, decreased lymphocyte proliferation and diminished antibody production. Deficiency in copper has been associated with increased susceptibility to infections.

It is important to note that increased zinc intake can induce copper deficiency because zinc competes with copper in the gastrointestinal tract. Thus, caution must be taken if copper is in the formulation when consuming zinc supplements.

It should also be noted that long-term high intakes of copper should be carefully monitored because it can result in adverse effects on immune function.

3.7 Manganese

3.7.1 *Introduction*

Dietary sources of manganese include whole-grain cereals, nuts, legumes, green leafy vegetables, raw egg yolks, beets, peas, blueberries, apricots, wheat germ and tea. Deficiency of manganese is rare. Manganese is absorbed in the small intestine, transported with protein carriers and excreted in bile. Only ~10% of manganese is absorbed, unless need is high. Manganese is absorbed better if iron status is good.

3.7.2 *Functions of manganese*

Manganese shares functional similarities with zinc and copper. It is needed for normal immune function, energy metabolism and acts as a cofactor for enzymes. For instance, it is a cofactor for manganese SOD which is located in the mitochondria. In neonates receiving parenteral nutrition, excessive manganese may induce neurotoxicity.

Deficiencies are associated with depressed thyroid functions and can cause increased susceptibility to allergies.

3.8 Selenium

3.8.1 *Introduction*

Dietary sources of selenium include seafood, meats, cereal grains and nuts. Selenomethionine and selonocysteine are the two forms of selenium found in nuts, oily fish, yeasts, oysters, clams, wheat germ, mushrooms and whole grains (Figure 3.5). The level of selenium in plant-derived food is dependent on selenium content in the soils.

Selenium is an essential mineral and adequate intake is essential to mount a proper immune response. At Linxian province where selenium deficiency is common, hyperendemic rates of upper gastrointestinal cancers have been reported. The soil at Linxian province of China is selenium-deficient. This observation suggests that selenium can reduce the risk of upper gastrointestinal cancer.

It should be noted that laboratory-made selenomethionine, sodium selenite and amino acid chelates sold in shops are not recommended due to the reported side effects and absence of beneficial effects. However, bioactive selenium made from greenhouse mustard greens and activated selenium consisting of selenocysteine plus methionine, vitamin E, riboflavin, broccoli and garlic were found to be beneficial supplements.

3.8.2 *Functions of selenium*

Selenium is required for the function of several selenium-dependent enzymes known as selenoproteins (proteins that contain a seleno-cysteine amino acid residue). For example, the glutathione peroxidases are selenoproteins that function in balancing redox and as antioxidants. These roles have implications for immune function and reducing cancer risk.

Selenium is well-known for its antioxidant properties. Selenium not only protects the host from oxidative stress but also plays an important

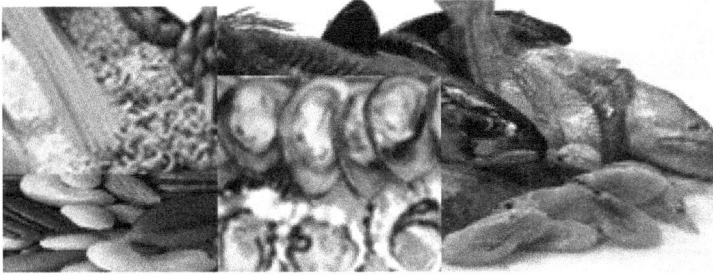

Figure 3.5 Examples of dietary sources of selenium.

Key points on selenium functions:

- Selenium acts as a cofactor in the antioxidant defense network (similar to vitamin E; protects cell membranes from oxidation)
- Selenium is involved in the synthesis of thyroid hormones
- Selenium may decrease risk of cancer (especially prostate and lung)
- Selenium is a cofactor in selenoproteins
- Lack of selenium deprives the cell of its ability to synthesise selenoproteins

  (Examples of selenoproteins are as follows:

  Iodothyronine deiodinase: produces active thyroid hormone (T3) from inactive precursor (T4)

  Thioredoxin reductase: reduces nucleotides in DNA synthesis and helps control intracellular redox state

  Glutathione peroxidase: protects body from oxidative damage. It catalyses this reaction:

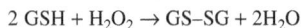

  $$2\ GSH + H_2O_2 \rightarrow GS\text{–}SG + 2H_2O$$

- Selenium has immune stimulating function. It is a component of the selenoproteins which regulate the function of immune cells, e.g. activation of natural killer cells, enhanced proliferation of T cells and regulation of cytokine expression.

---

role in balancing the redox state. The selenoenzyme thioredoxin reductase affects the redox regulation of several proteins including ribonucleotide reductase (an enzyme that reduces nucleotides in DNA synthesis).

Key points on immune effects of selenium deficiency:

- Enhances virulence of viruses such as enterovirus, coxsackie and influenza viruses
- Inhibits antibody production
- Impairs neutrophil chemotaxis

3.8.3 *Deficiency of selenium: Susceptibility to infection and cancer risk*

Selenium deficiency impairs both innate and adaptive immunity. Cell-mediated immunity is reduced.

The virulence of an enterovirus causing cardiac myopathy was associated with selenium deficiency in a region in Northeast to Southwest China. The selenium content in the diet is directly related with that of the soil in which the food was grown. Thus, selenium-deficient soils give rise to a deficit of this element in the population, as is the case in Northeast to Southwest China. Selenium deficiency can be a contributing factor to primary malnutrition in these areas where selenium content in soil is low. High incidence of Keshan disease was also reported in these regions. Patients suffer from Keshan disease which manifest as cardiac myopathy, which is due to enteroviral infection.

Experimental studies also suggest that selenium deficiency contributes to an increase in susceptibility to bacterial or viral infections like influenza and coxsackie viruses, and increases the risk of complications from infections. Selenium deficiency decreases antibody (IgM and IgG) production, impairs neutrophil chemotaxis, and has been shown to increase the virulence of coxsackie virus. Selenium supplementation can counteract the decrease in antibody production.

Two studies have shown that the immune response was stimulated in selenium-deficient normal or immune-suppressed individuals receiving 200 microgram per day of sodium selenite supplementation for eight weeks. Selenium was also found to regulate the expression of cytokines such as IL-2 and IL-8 in selenium-supplemented human immunodeficiency virus-infected patients.

It should be noted with caution that high concentrations of selenium can cause oxidative damage to immune cells.

Evidence for the role of selenium in reducing cancer risk is derived from a study in Linxian province in central north China where high incidences of upper gastrointestinal cancer were recorded. Soils in this province are selenium-deficient and cancer risk was mitigated with selenium supplementation in selenium-deficient individuals.

3.9 Iodine

3.9.1 *Dietary sources and absorption by the thyroid gland*

Dietary sources of iodine include saltwater seafood, seaweed, iodised salt, molasses and dairy products. Half a teaspoon of iodised salt meets the Recommended Dietary Allowance (RDA) requirement. Iodine is absorbed efficiently, stored in the thyroid gland and excreted in urine.

3.9.2 *Functions of iodine*

The synthesis of thyroid hormones thyroxine (T4) and triiodothyronine (T3) requires iodine.

Deficiency in iodine can result in lower thyroid activity, lower body temperature and cretinism (deafness-mutism, poor growth, mental deficiency). Deficiency is also associated with reduced microbicidal activity of leukocytes.

Excess iodine can result in enlarged thyroid gland and increased risk of thyroid cancer. In Japan, high intake of seaweed can result in excess iodine levels. In Chile, high levels of iodine are found in soils.

3.10 Chromium

3.10.1 *Dietary sources*

Chromium is found in many foods including broccoli, processed meats, liver, eggs, and whole grain products.

3.10.2 *Functions of chromium*

Chromium may improve insulin function and help normalise blood sugar.

3.11 Fluoride

3.11.1 *Dietary sources and functions*

The food sources of fluoride include fluoridated water, tea, seafood and seaweed. Fluoride is easily absorbed and stored in teeth and bones. The function of fluoride is to promote bone and dental health.

3.12 Other trace minerals

Trace minerals that are needed in very small amounts but essential for many enzymes are molybdenum, nickel, silica and boron. Molybdenum serves as a cofactor for enzymes such as sulphite oxidase, xanthine oxidase and aldehyde oxidase which are enzymes that are involved in the catabolism of sulphur amino acids and heterocyclic compounds. Legumes, grains and nuts are sources of molybdenum. Deficiency has not been seen in healthy people. Omission of molybdenum in long-term parentally fed patients resulted in amino acid intolerance.

The specific functions of nickel are not known but levels of nickel were increased in patients following heart attacks, burns, and strokes. Decreased levels of nickel have been seen in patients with hypotension, psoriasis and liver cirrhosis. Nickel deficiency has not been reported in humans but animal studies show that deficiency can cause dermatitis, pigment changes, decreased growth and reproductive capacities, and compromised liver function. Patients with severe nickel poisoning develop pulmonary and gastrointestinal toxicity. Sources of nickel include whole grains, nuts, beans and seafood.

Food sources of silica include oats, millet, barley, potatoes and horsetail herb. Silica is essential for smooth skin, shiny hair, beautiful nails, structure to stand upright, and strong bones. Silica has been shown to bind pathogens, activate phagocytes, reduce IgG and improve natural killer cell function. Boron, like silica, plays a role in preventing osteoporosis.

Summary of Chapter 3

Micronutrient intake influences antioxidant defences because trace elements are present in metallothionein (zinc), caeruloplasmin (copper), catalase (iron), SODs (copper, selenium, zinc), and glutathione peroxidase (selenium). SOD converts superoxide radicals to hydrogen peroxide and oxygen. The mitochondrial SOD contains manganese and the cytosolic SOD contains copper and zinc. Catalase converts hydrogen peroxide into oxygen and water, and glutathione peroxidase converts hydrogen peroxide to oxidised glutathione and water. Catalase contains iron while glutathione peroxidase contains selenium. Deficiency in zinc and selenium can contribute to increased susceptibility to infectious diseases. The most profound effect of copper deficiency on the innate immune system is the reduced number of neutrophils and decreased antibody production. Clinical signs of copper deficiency are anaemia, neutropenia, depressed growth, and abnormal bone development. Selenium has a possible role in reducing cancer risk. Selenium supplementation to selenium-deficient individuals living in Linxian province in central north China, where soils are deficient in selenium, mitigated the risk of upper gastrointestinal cancer. Iodine is required for synthesis of thyroid hormones and deficiency results in lower thyroid activity. Chromium may improve insulin function and help to normalise blood sugar. The trace minerals silica and boron strengthen the bones. Molybdenum is a cofactor for enzymes such as xanthine oxidase which are involved in catabolism of sulfur amino acids and heterocyclic compounds. The role of nickel is not clear but decreased levels were found in patients who have hypotension, liver cirrhosis and psoriasis. Appropriate intake of minerals is required for proper functioning of the immune system.

Chapter 4

Fatty acids and inflammation

Learning objectives

After studying this chapter, you should be able to:

1. Explain the types of fatty acids and their classification system
2. Describe the beneficial effects of fatty acids
3. Describe how fatty acids are synthesised and metabolised
4. State examples of fatty acids which have pro-inflammatory effects
5. Describe how fatty acids can resolve inflammation
6. State the sources of resolvers of inflammation
7. Explain the immunomodulatory roles of fatty acids

4.1 Introduction

Dietary fat (lipid) is mainly in the form of triglycerides, which are esters of glycerol and free fatty acids. Fats occur naturally in food. They are used to store energy in the body, insulate body tissues, cushion internal organs, and transport fat-soluble vitamins in the blood. Body fats are found as components of membrane phospholipids and cholesterol, which is a precursor of steroid hormones and bile salts required for fat digestion. The digestion of dietary fat gives rise to fatty acids. In blood, fatty acids are bound to albumin. Body fatty acids are also part of membrane phospholipids, which are precursors for synthesis of bioactive lipid mediators such as prostaglandins, leukotrienes, lipoxin, protectins and resolvins.

> **Key points on body fatty acids:**
> - Found in blood bound to albumin
> - Linked to glycerol to form triglycerides
> - Form part of membrane phospholipids
> - Precursors for synthesis of bioactive lipid mediators such as prostaglandins, leukotrienes, lipoxin, protectins and resolvins

4.2 Types of fats

There are two types of fats, namely, saturated and unsaturated fats. The source of saturated fats is mainly animal fats. Unsaturated fats can be classified under two types, namely, polyunsaturated and monounsaturated fats. The digestion of dietary fat gives rise to fatty acids, which are long hydrocarbons chains with a carboxyl group. Fatty acids vary in chain length and saturation, with the basic formula $CH_3(CH_2)_nCOOH$. Short fatty acids have less than 8 carbon atoms. Medium fatty acids have 8–14 carbon atoms. Long fatty acids are longer than 14 carbon atoms.

4.3 Structure of fats

Fatty acids are either saturated or unsaturated. Saturated fatty acids do not contain double bond(s) in the carbon skeleton structure; unsaturated fatty acids do. Saturated fatty acids can be derived from consumption of animal fats and are preferentially stored in adipose tissue. Examples of saturated fatty acids are palmitic acid and stearic acid (Figure 4.1), which are the major saturated fatty acids in the Western diet.

Examples of unsaturated fatty acids are oleic acid (OA), linoleic acid (LA), eicosapentaenoic acid (EPA) and docosahexaenoic acid (DHA), shown in Figure 4.2.

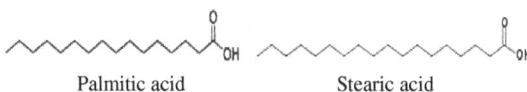

Palmitic acid   Stearic acid

Figure 4.1   Structure of palmitic and stearic acid. Palmitic acid has 16 carbons and stearic acid has 18 carbons, both with a carboxyl group at the terminal carbon.

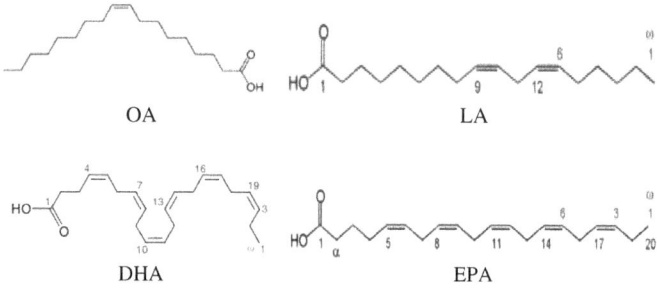

Figure 4.2   Structure of unsaturated fatty acids. OA has 18 carbons and is a monosaturated fatty acid as it has one double bond (C18:1). LA is 18 carbons long and has 2 double bonds (C18:2). DHA has 22 carbons and 6 double bonds (C22:6). EPA has 20 carbons and 5 double bonds (C20:5).

What is the difference in chemical structure between monounsaturated versus polyunsaturated fatty acids (PUFAs)? The difference is that monounsaturatured fatty acids contain one double bond and PUFAs contain more than one double bond, as shown below.

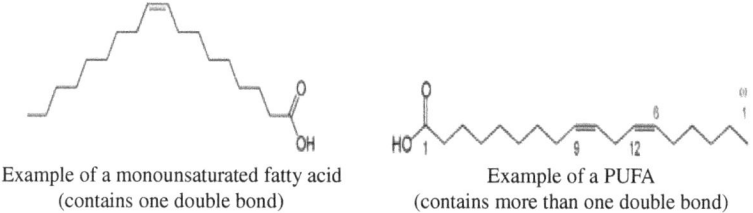

Example of a monounsaturated fatty acid
(contains one double bond)

Example of a PUFA
(contains more than one double bond)

4.4 *Cis* and *trans* fatty acids: Differences

Cis means the carbon groups are on the same side of a double bond, while *trans* indicates they are on opposite sides. The *cis* configuration prevents tight packing of fatty acids in membranes and hence increases membrane fluidity. *Trans* fatty acids (e.g. margarine) tend to be solids at room temperature, whereas *cis* fatty acids (e.g. vegetable oils) tend to be liquids. By artificially hydrogenating vegetable oils, the number of double bonds is reduced which causes the formation of *trans* fatty acids. Hydrogenation yields food products that are less vulnerable to rancidity but contain *trans* fatty acids. Not all *trans* fatty acids in the diet are due to food processing. For example, natural butter contains 5% *trans* fat.

4.5 Why is consumption of *trans* fatty acids not beneficial?

Increased consumption of hydrogenated vegetable oil in margarines leads to increased *trans* fatty acid consumption. When *trans* fatty acids are incorporated into cell membranes, the membrane fluidity is reduced and the cells do not function as well. *Trans* fatty acids behave as if they were saturated fatty acids, increasing circulating low-density lipoprotein and decreasing high-density lipoprotein cholesterol concentrations, which in turn raise the risk of cardiovascular disease. In most countries, nutrition labels for all conventional foods and supplements must indicate the *trans* fatty acid content.

4.6 Why is high fat intake not beneficial?

Epidemiological studies have found that a high fat intake is associated with a high risk of developing obesity, cardiovascular disease, type II

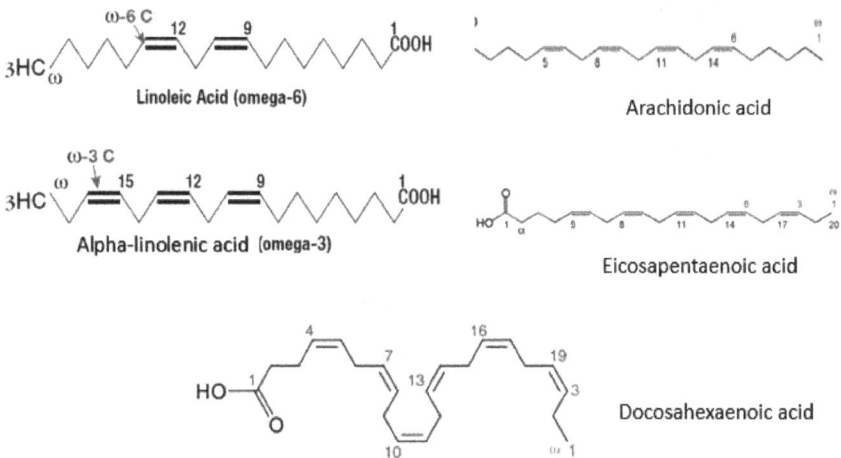

Figure 4.3 Examples of structures of ω-3 (n-3) or ω-6 (n-6) PUFAs.

diabetes and cancer (e.g. colon, prostate and breast). It is often suggested that the consumption of monounsaturated fatty acids, such as those in olive oil (the 'Mediterranean diet'), or ω-3 PUFAs should be increased accompanied by reduction in saturated fatty acids.

4.7 Classification of PUFAs as ω-3 or ω-6

The location of first double bond nearest to the terminal methyl carbon denotes whether the PUFA is ω-3 (also referred to as n-3) or ω-6 (also referred to as n-6). The examples shown in Figure 4.3 include LA and arachidonic acid, which are ω-6 PUFAs as the first double bond from the terminal methyl carbon is located at carbon position 6. Alpha-LA (ALA) is an ω-3 PUFA as the first double bond from the methyl end of the fatty acid is located at the third carbon atom. The structure of ALA is referred to as C18:3 n-3 as it has 18 carbons, 3 double bonds with the first double bond at the third carbon atom from the methyl end. Other examples of ω-3 PUFAs are EPA, which has a structure of C20:5 n-3, and DHA [C22:6 n-3].

4.8 Dietary sources of monounsaturated and polyunsaturated fatty acids

The major dietary monounsaturated fatty acid is OA, which is an ω-9 fatty acid and can be synthesised in the body. Dietary PUFAs are derived from plants consisting predominantly of ω-6 and ω-3 LA. DHA and EPA, which are also referred to as long chain ω-3 PUFAs, can be found in oily fish and fish oil. The table below summarises examples of food that are rich sources of these fatty acids (Table 4.1).

Table 4.1   Examples of fatty acids and dietary sources.

Example	Category	Structure	Rich Sources
OA	ω-9	C18:1 n-9	Olive oil, peanut oil, soy oil
LA	ω-6	C18:2 n-6	Canola, flaxseed, walnut, soybean
ALA	ω-3	C18:3 n-3	Flaxseed
DHA	ω-3	C22:6 n-3	Fish oil
EPA	ω-3	C20:5 n-3	Fish oil

Essential fatty acids are defined as fatty acids that have to be derived from the diet and are essential to life. They cannot be synthesised by humans due to the lack of desaturase enzymes required for their production. LA and ALA are examples of essential fatty acids. LA is the major ω-6 PUFA and ALA is the major ω-3 PUFA. Usually, the dietary intake of LA is much greater (5–20 fold) than ALA. LA can be found in foods such as vegetable oils (safflower, corn, soybean, canola), nuts, chicken fat, lard, egg yolk, butter and cheese. Sources of ALA which can be converted to ω-3 PUFAs are found in plants including pumpkin seeds, flaxseeds, hemp seeds, chia seeds, and walnuts. Examples of oils high in ω-6 PUFAs include most vegetable oil blends (typically labelled "vegetable oil"), sesame, sunflower, corn, soy, safflower and soy oils. Lower amounts of ω-6 PUFAs are found in olive, avocado, peanut, or canola oils. A rich source of dietary ω-3 PUFAs such as DHA and EPA is derived from certain species of fish e.g. salmon, tuna. A traditional Mediterranean diet typically consists of large quantities of fresh fruits and vegetables, whole grains, olive oil, legumes, nuts, fish and grape wines coupled with moderate physical activity that has been associated with reduced risk of heart disease, certain cancers, diabetes, Parkinson's and Alzheimer's diseases. This diet is rich in protective substances, such as antioxidants (especially resveratrol from wine and polyphenols from olive oil), vitamins C and E, selenium, glutathione, high amounts of fibre and a balanced ratio of n-6:n-3 fatty acids. This content is thought to be beneficial as it lowers the risk of cancer development and has cardioprotective effects. In Southeast Asian countries such as Malaysia, the potential sources of dietary ω-3 PUFAs include fish ('jelawat', 'sikap', sardines, tuna, mackerel and salmon), seafoods (shrimps, crab), vegetables, soybean, beans, peas, soy-based products, and ω-3 fortified foods such as eggs, ready-to-drink ω-3 milk preparation and soybean milk. The ratio of ω-3:ω-6 PUFAs in the human 'native' diet (plant-based food) ranges from 1:1 to 1:4, as compared with 1:11 to 1:20 in a typical modern Western diet.

A typical modern Western diet consists of refined or processed carbohydrates, increased saturated fats and *trans* fatty acids, high in ω-6 fatty acids (ratio of ω-3:ω-6 fatty acids = 1:11 to 1:20) and absent or low in fruits and vegetables. Increased refined carbohydrates can lead to hyperglycemia. Increased need for glucose metabolism leads to release of free radicals which cause oxidative stress and tissue damage. *Trans* fatty acids increase pro-inflammatory cytokine such as tumour necrosis factor (TNF) and induce acute inflammatory response, and imbalance towards higher ω-6 fatty acids increases inflammation. Thus, high intake of a typical Western diet is likely to be unbeneficial.

The consumption of ω-3 PUFAs is widely accepted as beneficial in chronic inflammatory diseases. ω-6 PUFAs are precursors of inflammatory mediators such as prostaglandins and leukotrienes, and are generally considered as pro-inflammatory and contribute to the progression of chronic inflammation. An exception is primrose oil which is rich in ω-6 PUFAs but exerts anti-inflammatory effects. An imbalance of excessive ω-6 and deficiency of ω-3 PUFA consumption is associated with higher risk of cardiovascular mortality and other inflammatory diseases. This may be due to the difference in the lipid metabolites derived from LA and ALA. One way to increase ω-3:ω-6 ratio is to increase consumption of oily fish and fish oil, ω-3 rich foods such as flaxseed oil, and reduce intake of ω-6 rich vegetables oils. The next section shows how ALA can be metabolised in the body and then converted to ω-3 PUFAs whilst LA is converted to an ω-6 PUFA, arachidonic acid.

4.9 Metabolism of LA and ALA

LA from dietary intake is metabolised to arachidonic acid, whilst ALA is metabolised to EPA and DHA. The enzymatic steps involved are as shown in the figure below (Figure 4.4). ALA is metabolised to EPA by a desaturation and two elongation steps. Elongation of EPA gives rise to DHA.

4.10 Metabolism of arachidonic acid

Arachidonic acid is the major precursor of eicosanoids. It is found at higher levels in cell membranes compared to EPA. Arachidonic acid is metabolised by three main classes of enzymes: (i) cyclooxygenase generating prostaglandins and thromboxanes, (ii) lipoxygenase activity producing leukotrienes and lipoxins and (iii) p450 epoxygenases which synthesise epoxyeicosatrienoic acids (EETs) and HETEs (Figure 4.5).

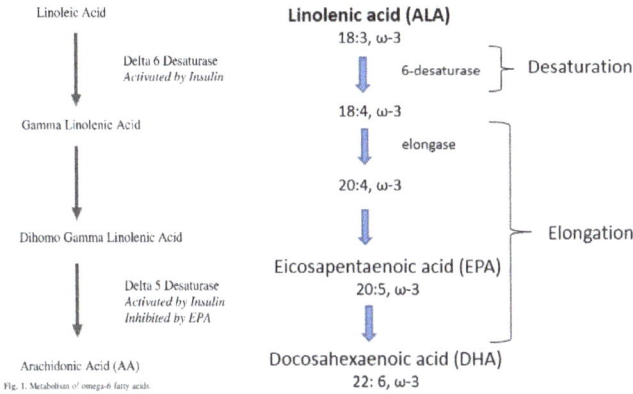

Figure 4.4   Metabolism of LA and ALA.

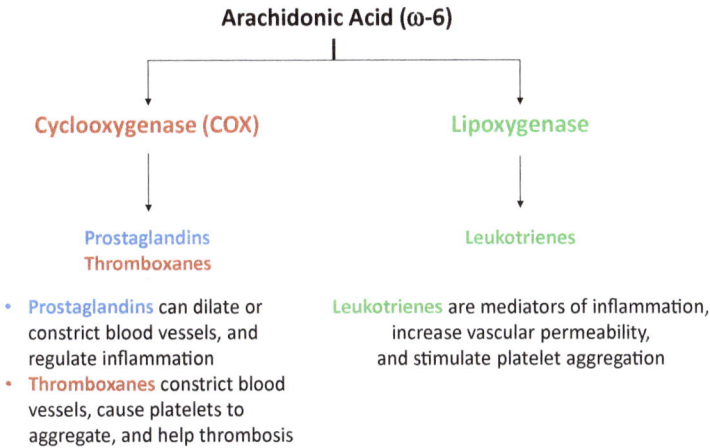

Figure 4.5   End-products of metabolism of arachidonic acid.

4.11   Metabolism of EPA and DHA to pro-resolving lipid mediators

EPA and DHA are the precursors for the production of anti-inflammatory lipid mediators such as resolvins and protectins (Figure 4.6), which are involved in the resolution of acute inflammation. Resolvins, protectins and lipoxins are also known as specialised pro-resolving lipid mediators (SPMs).

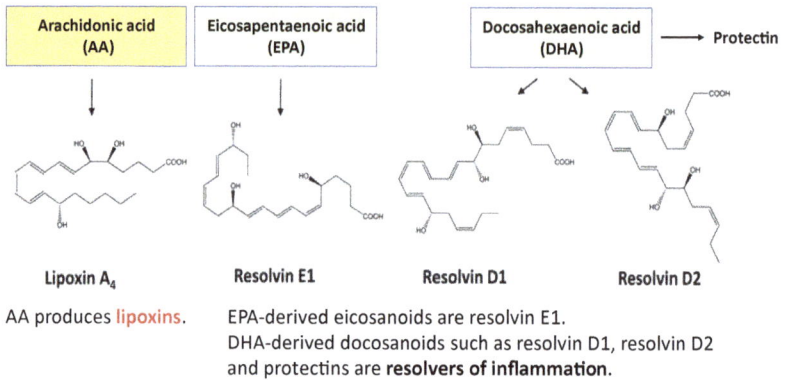

Figure 4.6   **Precursors of resolvers of inflammation, resolvins and protectins.** Arachidonic acid is the source of substrate for production of lipoxin. EPA is the substrate for production of resolvin E1. DHA produces protectins, resolvin D1 and resolvin D2.

4.12 Roles of SPMs in inflammation and the adaptive immune responses

Acute inflammation is generally protective and is mediated by many chemical factors including cytokines, chemokines and SPMs. In general, arachidonic acid-derived eicosanoids are lipid mediators with both potent pro-inflammatory and anti-inflammatory effects. Eicosanoids such as prostaglandin E2, thromboxane and leukotriene LTB_4 derived from arachidonic acid have pro-inflammatory effects. Prostaglandin PGE_2 and leukotriene LTB_4 are mediators of inflammation and they cause vasodilation (increased blood flow), which results in redness. LTB_4 increases the permeability of blood vessels, resulting in plasma leaking out into the connective tissues, and swelling occurs. Thus, prostaglandins and leukotrienes which are derived from arachidonic acid contribute to acute inflammation which is characterised by its cardinal signs, namely, rubor (redness), calor (heat), tumour (swelling) and dolor (pain). Typically, acute inflammation is meant to be transient. After killing of the invading microbes, the neutrophils undergo apoptosis and need to be cleared to allow homeostasis to be restored. This is facilitated by changes in lipid mediators (eicosanoids) and macrophage phenotype.

These changes contribute to resolve the inflammation and proceed to wound healing and repair. Lipid mediators switch from prostaglandin and leukotrienes to lipoxins and resolvins to promote the resolution of inflammation. Monocytes are recruited and neutrophil infiltration is reduced. Monocytes differentiate to M1 macrophages and switch to M2 macrophages which are important for resolution of inflammation and induction of tissue repair.

Key points on acute inflammation:
- Increased vasodilation (by histamine, prostaglandins, etc.); prostaglandins derived from metabolism of arachidonic acid regulate changes in blood flow
- Increased vascular permeability (by histamine, leukotrienes, etc.)
- Leakage of fluid containing proteins and cells such as neutrophils from blood into site of inflammation
- Inflammatory mediators attract neutrophils to site
- Neutrophils engulf harmful pathogen

The pro-resolving lipid mediators of inflammation are arachidonic acid-derived lipoxins, EPA-derived eicosanoids such as resolvin E1 and DHA-derived docosanoids such as resolvin D1, resolvin D2 and protectins. These lipid mediators play an important role in controlling the magnitude and extent of the inflammatory events, preventing inflammation from spreading, and stopping the transition from acute to chronic inflammation.

Lipoxins signal macrophages to perform efferocytosis, which is a process whereby dead cells and apoptotic neutrophils are phagocytosed (Figure 4.7). Efferocytosis of cell debris is then followed by exit via the lymphatic vessel, and the tissue can proceed to wound healing and repair. SPMs can also modulate the adaptive immune responses. Cytotoxic CD8$^+$ T cells kill infected, cancer and damaged cells through the release of granzymes and perforins. They also potentiate the function of innate immune cells through the release of cytokines such as IFN-γ and TNF. *In vitro* studies have shown that SPMs can dampen the release of cytokines from CD8$^+$ T cells and contribute indirectly to the resolution of inflammation by preventing the recruitment and activation of

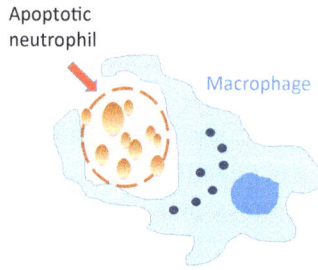

Figure 4.7   Efferocytosis of apoptotic neutrophils by macrophages.

innate immune cells, thus avoiding chronic inflammation and immune-mediated damage.

CD8$^+$ T cells can also prime and re-stimulate CD4$^+$ Th cells to release cytokines. The uncontrolled and persistence of T helper cells such as Th1 and Th17 are often associated with tissue damage and pathological states such as in autoimmune diseases. Resolvins can dampen Th1 and Th17 responses by preventing their generation. Resolvins can also enhance the differentiation of CD4$^+$ T cells to Tregs. Tregs can dampen the excessive immune responses and prevent the over-activation of Th1 and Th17 cells, thus reducing immune-mediated damage.

Key points on resolution of inflammation from *in vitro* or animal studies:

- Lipid mediators produced switch to lipoxins, protectins and resolvins which are also known as SPMs
- Monocytes are recruited and less neutrophils infiltrate into the site of inflammation
- Monocytes which had differentiated to M1 macrophages switch to M2 macrophages, which play a central role in resolving inflammation and inducing tissue repair
- SPMs inhibit the release of cytokines from cytotoxic CD8$^+$ T cells and prevent the recruitment and activation of innate immune cells
- Resolvins inhibit pro-inflammatory responses by Th1 and Th17 and enhance differentiation of CD4$^+$ T cells to Tregs, which prevent overactivation of the pro-inflammatory Th1 and Th17 cells

---

4.13 Modulation of the immune system by ω-3 PUFAs

ω-3 PUFAs can modulate the inflammatory response in neutrophils. Animal studies show that ω-3 fatty acids decrease activation of macrophages and neutrophils, production of reactive oxygen species, phagocytosis and antigen presentation (by macrophages). Animals fed with ω-3 PUFA-rich diets showed decreased interleukin (IL)-1β, IL-6 and IL-8 production, whereas ω-6 PUFA-rich diets increased the production of these cytokines. The decrease in expression of pro-inflammatory genes such as TNF and IL-1β by ω-3 PUFAs is via decreasing the activity of the transcription factor NFκβ. This transcription factor is considered as a pro-inflammatory nuclear factor.

The expression of adhesion molecule and lymphocyte adhesion to the endothelium was also decreased, thus reducing the recruitment of neutrophils to the site of inflammation. The resolvers of inflammation, resolvins and protectins, can promote neutrophil apoptosis and phagocytosis.

Low concentrations of ω-3 fatty acids are needed for lymphocyte proliferation, but at high concentrations, lymphocyte proliferation is inhibited. Antibody synthesis and IL-2 production is also inhibited. NK activity was decreased.

In summary, long chain ω-3 PUFAs exert anti-inflammatory effects by: (1) decreasing activation of macrophages and neutrophils. This will reduce the production of reactive oxygen species, phagocytosis, and chemotaxis in response to inflammatory stimuli; (2) reducing neutrophil adhesion to the vascular endothelium. This will reduce neutrophil transmigration across the endothelium and into the affected tissue; and (3) promoting the synthesis of resolvins and protectins. Figure 4.8 shows that the anti-inflammatory ecosanoids, lipoxin and resolvin E1 stop migration of neutrophil infiltration into the site of inflammation, decrease IL-12 production by APCs and decrease proinflammatory cytokines by Th1 cells. They also stimulate recruitment of monocytes into the site of inflammation and stimulate macrophages to perform efferocytosis (uptake of apoptotic neutrophils).

Animal studies have shown that consumption of a Western diet, which is typically higher in n-6 PUFAs, results in greater formation of arachidonic acid as compared to EPA in immune cell membranes. When intake of ω-3 PUFA acid is increased, EPA content in immune cell membranes increases due to increased production of eicosanoids derived from EPA such as resolvins, and decreased production of prostaglandins and leukotrienes derived from arachidonic acid, leading to an overall anti-inflammatory effect.

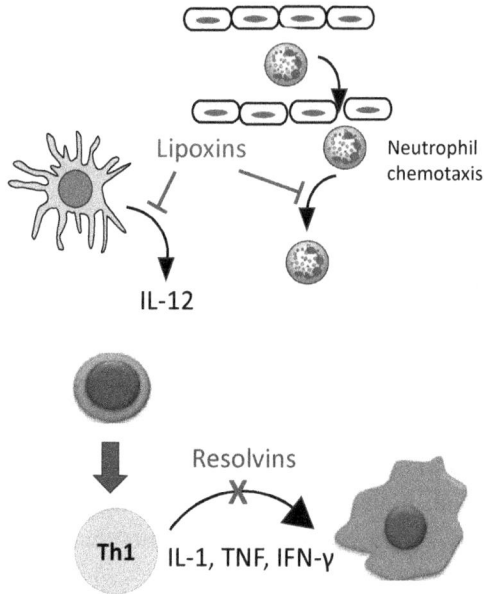

Figure 4.8 Effects of resolvins and lipoxins.

In vitro studies have shown that ω-3 PUFAs are incorporated into membrane phospholipids of immune cells. This alters the physical properties of the cell membrane, disrupts membrane organisation, and modulates immune responses, such as phagocytosis, MHC protein expression and T-cell signaling. Cholesterol and proteins are redistributed. Disruption of the peptide-MHC interaction with the T cell receptor at the membrane surface can occur. Altering the lipid rafts by ω-3 PUFAs changes the membrane protein composition and this will alter signal transduction networks, resulting in biological changes such as suppression of genes involved in T cell activation, NK cell activity, dendritic cell maturation and phagocyte function.

Key points on immunomodulatory effects of ω-3 PUFAs:

- Dendritic cell maturation and functions of phagocytes, T cells and NK cells are inhibited
- Neutrophil chemotaxis in response to inflammatory stimuli are decreased
- Proinflammatory cytokines levels are decreased

- Neutrophil adhesion to the vascular endothelium and neutrophil transmigration across the endothelium and into the affected tissue are reduced
- Synthesis of resolvins and protectins is promoted
- Physical properties of the cell membrane such as organisation are disrupted, and immune responses such as phagocytosis, MHC protein expression and T-cell signaling are altered

4.14 Saturated fatty acids and inflammation

Diets rich in saturated fat have been long been thought to contribute to conditions associated with dyslipidaemia and atherosclerosis. Mice fed with a diet high in fat over 4 weeks were found to have an increased expression of inflammatory cytokines as well as elevated circulating lipopolysaccharides which act as endotoxins. In mice, it has been shown that endotoxins can be absorbed from the intestine and transported to the blood circulation. Healthy human subjects fed with a high-fat meal also show high circulating concentrations of lipopolysaccharides. Current knowledge suggests that the source of endotoxin is derived primarily from gut microbes and also from foods such as meats, cheese, and dairy products, since lipopolysaccharide is the ligand that binds Toll-like receptor 4 resulting in expression of inflammatory genes such as IL-1β, TNF and COX-2 via activation.

4.15 Dietary fat, Mediterranean diet and health

Epidemiological studies have found that diets high in saturated fat and *trans* fat are pro-inflammatory in nature. It is likely that this is due to the ability of saturated fats to promote absorption of endotoxin from gut microflora as shown by *in vitro* and animal studies. For many decades, it has been known that LA reduces blood cholesterol concentration and substituting saturated fats with LA lowers the risk of heart disease. However, concern was expressed that LA-rich diets are unhealthy and promote inflammation due to the production of arachidonic acid, leading to enhanced synthesis of prostaglandins and leukotrienes. Fortunately, human clinical trials failed to support the theory that LA promotes inflammation.

Key points on beneficial effects of increased dietary ω-3-PUFAs:

- Outcomes for patients with hypertriglyceridemia, hypertension and secondary cardiovascular disease prevention are improved
- Primary prevention of rheumatoid arthritis and cardiovascular diseases

---

There is strong and consistent evidence that consumption of a Mediterranean diet may help reduce inflammation and lower the risk of cardiovascular disease. A Mediterranean diet is rich in fruit and vegetables, olive oil, nuts, beans, fish, and whole grains which are good sources of essential fatty acids, namely, ALA, EPA and DHA. Higher intake of these PUFAs has been associated with decreased levels of clinical markers of inflammation, namely, C-reactive proteins and serum amyloid protein A. EPA and DHA serve as precursors for resolvins, which are resolvers of inflammation. They stimulate endothelial production of nitric oxide and vasoprotective prostacyclin which act as anti-adhesive and anti-thrombotic mediators, thus contributing to reduced risk of cardiovascular disease.

Key points on DHA:

- Retinal cell membrane contains high concentrations of DHA
- Normal development and function of the retina requires DHA
- Accumulation of DHA in the brain and retina is most critical in the last trimester of pregnancy
- Insufficient DHA in preterm infants exposes them to harm due to slowing down of visual and neural development
- Visual function is significantly improved in preterm infants fed with DHA-fortified formulas

---

Eskimos and Inuit populations in Greenland and Alaska consume a diet rich in oily fish, which is the best characterised source of long chain ω-3-PUFAs, and other sources of lipids such as whale and seal meats. Epidemiological studies have shown that these populations are at lower risk of cardiovascular disease. However, it was also observed that the incidence of infectious disease such as tuberculosis is increased in these populations. This is consistent with animal studies showing that feeding animals with a diet rich in long chain ω-3-PUFAs resulted in susceptibility to *Mycobacterium tuberculosis*. Hence, while ω-3 PUFA supplementation may be beneficial to individuals with autoimmune or inflammatory diseases, caution needs to be taken as high ω-3 PUFA intake could possibly be harmful, as host-defense mechanisms are reduced and vulnerability to infectious diseases is increased.

4.16 Is supplementation with ω-3 PUFAs beneficial for patients with chronic inflammatory conditions?

It is rational to imply that if sufficient amounts of ω-3 PUFAs such as EPA are consumed, the concentration of arachidonic acid diminishes and consequentially pro-inflammatory eicosanoids are reduced. This effect may be beneficial in chronic inflammatory diseases because it results in reduced production of pro-inflammatory cytokines which include interferons, IL-1 TNF, and reactive oxygen species. It is implied that lower concentrations of reactive oxygen species can reduce tissue damage and dampen pathological inflammatory reactions. The benefits of n-3 PUFAs have been reported in some studies.

Studies have shown that consumption of ω-3 PUFAs in fish oil by patients with chronic inflammatory conditions such as rheumatoid arthritis is beneficial. Synovial fluid from joints of rheumatoid arthritis patients contains increased levels of pro-inflammatory cytokines such as TNF and IL-1β. Consumption of oily fish and fish oils containing EPA and DHA has revealed some clinical improvement as shown by reduced joint rigidity and pain, such that dosage of drugs such as non-steroidal anti-inflammatories can be reduced.

In theory, consumption of fish oil or ω-3 PUFAs by patients with inflammatory diseases such as rheumatoid arthritis, psoriasis, multiple sclerosis, ulcerative colitis, and Crohn's disease will be beneficial. However, there is limited evidence for the beneficial effects of fish oil or ω-3 PUFAs with the exception of rheumatoid arthritis. This could be due to the high doses of fish oil supplementation used in patients with rheumatoid arthritis.

Another group of chronic inflammatory state is cardiovascular disease, metabolic syndrome and type II diabetes mellitus. Obesity and abdominal obesity (also called visceral obesity) are risk factors for these diseases. The Western diet is characterised as a diet rich in saturated fats, processed meat and simple carbohydrates. It has been associated with adverse health outcomes, including metabolic syndrome, obesity and type II diabetes mellitus. Dietary fat has been proposed to drive chronic low-grade inflammation. Murine studies show that adipose tissues secrete pro-inflammatory factors known as adipokines, which promote macrophage recruitment to enlarged adipose tissues resulting in the generation of adipose inflammation. This explains the macrophage infiltration in adipose tissues of obese individuals. The expanded adipose tissues also release fatty acids, resulting in decreased expression of the glucose transport protein GLUT4 on muscle cell surfaces. Decreased GLUT4 leads to inhibition of both glucose uptake and glycogen synthesis during the fed state, leading to high blood glucose and development of type II diabetes mellitus. However, a longitudinal cohort study in Finland showed that increased serum ω-6:ω-3 PUFA ratio resulted in reduced risk of metabolic syndrome. The disparities between murine and human studies is unclear. Although murine studies showed beneficial effects of n-3 PUFAs, it is still unclear whether it is of clinical benefit for patients with metabolic syndrome and type II diabetes.

The clinical benefits of n-3 PUFAs for cardiocardiovascular disease are still unclear. Some studies have reported that supplementation with n-3 PUFAs resulted in 20–30% reduction in mortality from cardiovascular diseases. Beneficial effects of administering fish oils (EPA and DHA) in the form of capsules were reported for the Gruppo Italiano per lo Studio della Sopravvivenza nell'Infarto Miocardic (GISSI) Prevention Trial, which followed over 11,000 patients for 3.5 years after myocardial infarction. However, another study did not find beneficial effects of n-3 PUFAs on cardiovascular events in patients with glucose intolerance/diabetes and cardiovascular risk factors. It appears that the effects of n-3 PUFAs in cardiovascular disease are more complicated.

Recent studies in animal models have shown that the gut microbiota, which resides mostly in the lower intestine, plays a role in generating inflammation and regulating obesity and metabolism. The microbiota consists of trillions of microbes living in the gut and can be altered in response to micronutrients. Animal studies have shown that n-6 PUFA causes dysbiosis (microbiota imbalance), inflammation, and metabolic disease. However, some human studies have shown that consumption of n-6 PUFA is not harmful. The influence of gut microbiota on the effects of consumption of n-6 and n-3 PUFA is still unclear.

Summary of Chapter 4

The classification of fatty acids into saturated and unsaturated fatty acids and chemical structure differences were described. The metabolism of two essential fatty acids, namely, LA and ALA, from dietary sources yields arachidonic acid, and EPA and DHA, respectively. Fish oil provides an abundant source of EPA and DHA. Arachidonic acid is further metabolised to prostaglandins and leukotrienes which are pro-inflammatory lipids. Lipoxin, which is a resolver of inflammation, is derived from arachidonic acid. EPA and DHA are precursors for resolvers of inflammation, resolvins and protectins. These lipid-derived mediators which can resolve inflammation are also known as SPMs. ω-3 PUFAs inhibit production of inflammatory mediators including proinflammatory cytokines (IL-1β, TNF-α, IL-6), chemokines (IL-8, MCP-1), eicosanoids (PGE$_2$, leukotrienes), and reactive oxygen and nitrogen species. Resolvins and lipoxin can stop recruitment of neutrophils to the site of inflammation, downregulate pro-inflammatory cytokine production and assist in the removal of dead cells, leading to wound healing and tissue repair. *In vitro* studies have shown that SPMs can also dampen inflammatory responses by reducing release of cytokines from cytotoxic CD8$^+$ T cells, resulting in reduced stimulation of CD4$^+$ T cells. SPMs can also enhance the differentiation of CD4$^+$ T cells to Tregs, which can then prevent the activation of pro-inflammatory Th1 and Th17 responses, leading to dampening of immune-mediated damage. The benefits of fish oil supplementation in inflammatory diseases such as cardiovascular disease, inflammatory bowel disease and cancer are still inconclusive as studies are inconsistent, although reduction in joint stiffness and pain have been observed in rheumatoid arthritis patients. The traditional Mediterranean diet which typically consists of green vegetables, olive oil, whole grains, legumes, nuts, fruits, fish and grape wines have been shown to be beneficial for prevention of cardiovascular disease. Long chain ω-3 PUFAs have anti-inflammatory effects and are deemed beneficial to reduce chronic inflammation. In rheumatoid arthritis patients, consumption of oily fish and fish oils containing EPA and DHA has revealed some clinical improvement as shown by the reduced joint rigidity and pain, such that dosage of drugs such as non-steroidal anti-inflammatories can be reduced. The anti-inflammatory effects of ω-3 PUFAs are conferred by their ability to inhibit a range of immune functions including suppression of T cell function, inhibition of antigen presentation, adhesion molecule expression, pro-inflammatory cytokine expression and eicosanoid production. However, high ω-3 PUFA intakes could possibly be harmful as they can weaken host-defense mechanisms and increase susceptibility to infectious diseases. Thus, caution needs to be taken as to whether ω-3 PUFA supplementation is beneficial.

Chapter 5

Food allergy

Learning objectives

At the end of studying this chapter, you should be able to:

1. Define the difference between food allergy and food intolerance
2. Describe the immune reactions that result in the symptoms associated with food allergy
3. Explain factors that contribute to food allergy
4. Present an overview of the methods used for diagnosis of food allergies and how to prevent and manage them

5.1 Introduction

Food allergy is defined as an adverse response to food that is immune-mediated. Examples of allergenic foods include milk and milk products, shellfish, cereals, egg and egg products, fish and fish products, legumes, peas, soybeans, peanut, gluten, tree nuts like brazil nuts, hazelnuts, walnuts, and their products.

Table 5.1 shows examples of allergens present in common foods. Chemicals, additives and dyes added in processed foods are factors that have contributed to increase in food allergies. Another factor is the cross-reactivity between allergens in the air and allergens in food. For example, cross-reactivity between birch pollen in the air and molecules in fruits (apple, peach, cherry, almond, plum, apricot and strawberry)

Table 5.1   Common allergenic foods and examples of allergens.

Cow's milk — α-lactalbumin, β-lactoglobulin
Egg — ovomucoid, ovalbumin
Shellfish e.g. shrimps — tropomyosin
Fish — allergen M in cod fish
Wheat — gluten
Peanuts — vinculin
Soybean — P22-25, P34 (thiol protease)
Tree nuts e.g. pistachio, brazil nuts, hazelnuts

and vegetables (carrot, potato, hazelnuts, fennel, celery) can confer food allergy.

5.2   Mechanisms of allergic response to food allergens

The mechanisms of immune-mediated reaction to foods are similar to immune response to a foreign antigen. The most common type of allergic response is mediated by IgE and is referred to as IgE-mediated or Type I hypersensitivity reaction. In a pre-sensitised individual, binding of food antigen to IgE prebound to the IgE receptor on mast cells causes cross-linking of IgE and activates degranulation of mast cells and release of chemical mediators such as histamine, prostaglandin and leukotrienes (Figure 5.1). The manifestation of IgE-mediated allergic reactions to food in the gastrointestinal tract include abdominal pain, nausea, vomiting, diarrhea and oral allergy syndromes (itching, swelling of lips, tongue and larynx, urticarial, rhinitis, asthma, laryngeal oedema and even anaphylaxis). Ingested food allergens can be transported to other parts of the body and affect the skin, respiratory tract and other systemic manifestations. Anaphylaxis is a serious systemic manifestation which can be fatal. It is manifested as generalised severe rapid reaction with itching and swelling of oral cavity, skin, respiratory, gastrointestinal and cardiovascular symptoms.

Symptoms associated with IgE-mediated reactions at the skin include urticaria (appearance of wheals in the skin), angioedema (swelling of subcutaneous tissue) and morbilliform rashes. In mixed IgE and

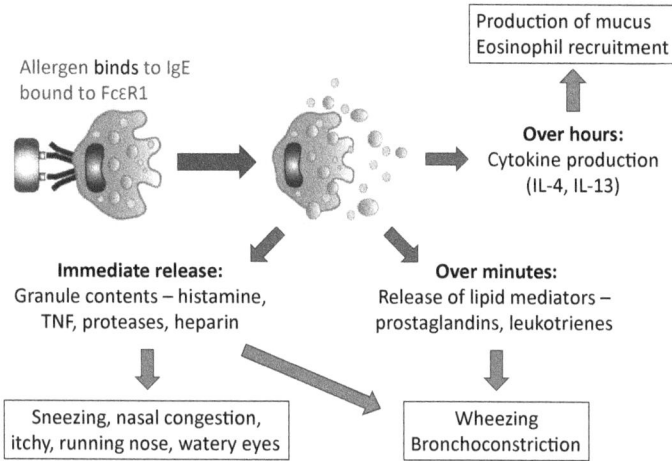

Figure 5.1 Immune response to allergens.

cell-mediated reactions, cutaneous reaction is manifested as atopic dermatitis (itching and drying skin).

Symptoms associated with IgE-mediated reactions at the respiratory tract include rhinitis (inflammation of nasal passages, runny nose), laryngeal oedema (constriction of the throat), wheezing and bronchospasm (narrowing of airways and breathlessness).

5.3 Food intolerance

Food intolerance is an adverse reaction to either the chemical or toxin in the food or it can be due to enzyme deficiency in the host. It does not have an immunological basis.

Food intolerance can be due to enzymatic food intolerance e.g. lactose intolerance or pharmacological food intolerance (e.g. food additives, biogenic amines such as histamine and canned food). In some cases, the mechanism underlying the adverse reaction is undefined.

Key points on types of food intolerance:

- Enzymatic intolerance
- Pharmacological
- Undefined

5.4 Factors that contribute to food allergy: Atopy

Atopy is defined as a genetic predisposition to mounting an IgE antibody response. Atopy is associated with allergic diseases such as allergic rhinitis, asthma, atopic dermatitis and also food allergies. The risk of developing IgE-mediated food allergy when one parent is atopic is 13.5–58% and increases up to 100% when both parents are atopic (European Commission 1997).

Key points on factors that influence susceptibility to food allergens:

* Atopy
* Early introduction to foods
* Level of consumption
* Food processing techniques
* Novel food and genetically modified foods

5.5 Early introduction to foods

The gut mucosal barrier takes at least approximately 10 years to fully mature. Greater permeability of the gut mucosa of young children allows more proteins including food allergens to enter the immune system and result in IgE-mediated food allergy.

5.6 Level of consumption

High consumption of certain foods such as peanut and fish may increase the likelihood of development of allergy to these foods.

5.7 Novel foods

The introduction of new food into the diet has been documented to contribute to development of food allergy. In the mid-1980s, soybean was introduced into the French diet and resulted in an increase in the incidence of soybean allergy. Genetically modified soybean containing a gene isolated from Brazilian nut consumed by individuals known to be allergic to it resulted in an allergic reaction to the genetically modified soybean.

There is also the possibility that genetically modified foods may harbour components that are not known to contain allergens.

5.8 Effect of heat, digestion and food processing techniques on allergenicity

In general, food allergens are heat-stable and are not destroyed at high temperatures. Allergens in fish, cow's milk and soybean are known to be heat-resistant and also resistant to enzymatic digestion by gastrointestinal secretions. However, hidden epitopes can be exposed by the process of digestion. The allergenicity of allergens in peanuts can be enhanced by heating and roasting. It has been reported that roasting of peanuts, which is a common practice in USA, increases the allergenicity of peanut proteins. This is in contrast to boiling or frying of peanuts, which is a common method used to cook peanuts in China where peanut consumption is high, but where prevalence of peanut allergy is lower than in USA.

Food processing techniques such as lyophilisation of fish allergens and phenolic browning of apple allergens can also diminish the IgE-binding capacities. Hence, food processing techniques may be a factor contributing to food allergy and safety issues.

5.9 Diagnosis of food allergy

5.9.1 *Medical history and physical examination*

The first step in the diagnosis of food allergy is to take a thorough medical history including a review of the family history to identify atopic individuals and to conduct a physical examination.

Diagnosis of food allergy:

- Take a medical history including a description of symptoms, family history and conduct a physical examination
- Keep a food diary
- Eliminate suspected food from diet
- Perform skin prick test

- Measure specific IgE in blood
- Perform oral food challenge

5.9.2 *Food diary*

Keeping a food diary of eating habits, symptoms and medications can help to pinpoint the problem.

5.9.3 *Elimination diet*

The suspected food is eliminated from the diet for a week or two and then added back into the diet. This method is not safe and should not be practised without immediate access to an emergency doctor or medical clinic facility if the individual had a severe reaction to the food in the past.

5.9.4 *Skin prick test*

To perform the skin prick test, a small amount of suspect food is pricked with a needle to embed the food substance beneath the skin of the forearm or back. A wheal and flare reaction appears in 15–25 minutes if the response is positive and disappears in 1–2 hours. This method is used to screen for food allergy but it should be noted that varied responses can be obtained from commercial extracts used in the skin prick test as opposed to the fresh food allergens.

5.9.5 *Blood test*

A number of allergens have been identified and blood test to measure the presence of IgE specific to these allergens have been developed.

5.9.6 *Oral food challenge*

In the clinic, the patient is given small but increasing amounts of the suspect food. This test needs to be carried out under medical supervision as there is a possibility of severe reaction to the food.

5.10 Prevention and management of food allergy prevention

Avoidance of allergenic food is a good preventive measure. Eliminating common allergenic food during pregnancy and lactation for women with a family history of allergy is beneficial. Delayed introduction of solid foods to an infant until after 4 months and limiting variety until at least 6 months is recommended. Breastfeeding has been reported to provide partial protection against allergy and this could be due to reduced exposure to proteins in infant formulas. The prolonged period of breastfeeding also helps in preventing allergy as this allows more time for maturation of the gut mucosal barrier of the infant.

5.11 Management and treatment of food allergy

Complete avoidance of the offending food is a good way to manage food allergy and ensure that the nutritional requirements are being met.

Summary of Chapter 5

This chapter explains the difference between food allergen and food intolerance. The mechanism underlying IgE-mediated hypersensitivity reaction to allergens and how it contributes to the symptoms of allergy was described. The factors that contribute to food allergy including atopy, level of consumption, introduction of novel foods and food processing techniques were elaborated. The diagnosis and management of allergy were also explained.

Chapter 6

Probiotics and prebiotics

Learning objectives

At the end of studying this chapter, you will be able to:

1. Explain the importance of gut microbiota in maintaining health
2. Explain the benefits of probiotics and prebiotics
3. Describe the effects of probiotics and prebiotics on immune cells

6.1 Introduction to gut microbiota

The gut microbiota is a community of microorganisms that live in the digestive tract of humans and animals. They are generally non-pathogenic and play an important role in maintaining health by providing nutrients and a protective barrier against invading pathogens. Examples of these commensal microorganisms are those belonging to the genus *Bacteriodes*, *Prevotella* and *Ruminococcus*. Other commensals such as acidophilus (e.g. *Lactobacillus acidophilus*) and bifido bacteria (e.g. *Bacillus Longum* and *Bacillus infantis*) have been found to support mucosal immunity by producing butyrate and other short chain fatty acids (SCFAs). Butyric acid produced promotes intestinal mucin, which acts like a lubricant to mobilise food through the intestine and prevent constipation. In addition, acidophilus bacteria help maintain an acidic environment, which can prevent the growth of harmful bacteria. Furthermore, bifido bacteria *Bacillus Longum* has been found to increase IgA on mucus membrane, suppress

IgE production, and increase interleukin (IL)-12 and interferon (IFN)-γ expression in mice. *Lactobacillus plantanum* can also increase expression of IL-12 and IFN-γ, which are known Th1 cytokines that support cell-mediated immunity. Diet has a great influence on microbial colonisation and the profile of major species in the human gut microbiota can potentially be modified by dietary intake. The two most abundant phyla found in most healthy individuals are the *Bacteroidetes* and *Firmicutes*.

6.2 Probiotics

6.2.1 *Introduction*

Probiotics are defined by the World Health Organisation as *"live microorganisms that can provide benefits to human health when administered in adequate amounts, which confer a beneficial health effect on the host"* (WHO/2001). The common probiotic microorganisms are *Bifidobacteria* and *Lactobacilli*. These probiotics are consumed in yogurt, cheese and other fermented foods such as kim chi, miso and fermented milk (Figure 6.1).

6.2.2 *Benefits of probiotics*

The benefits of *Bifidobacteria* was established more than 100 years ago by Tissier, who observed that gut microbiota from healthy breast-fed infants were dominated by *Bifidobacteria* but they were not present in formula-fed infants suffering from diarrhoea. Other studies have shown the clinical benefit of probiotic microorganisms such as *Lactobacillus* in decreasing the duration of diarrhoea. The frequency of antibiotic-associated diarrhoea is

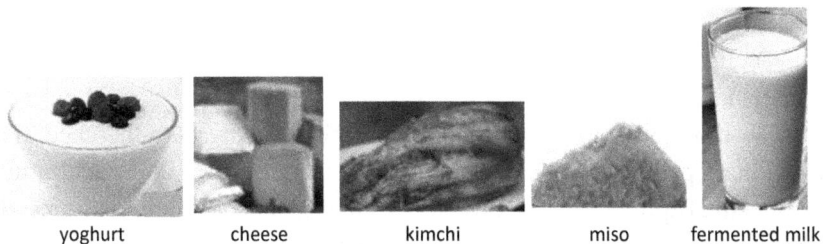

| yoghurt | cheese | kimchi | miso | fermented milk |

Figure 6.1 Examples of sources of probiotics.

also reduced in infants and the elderly when fed with diet which are supplemented with *Bifidobacterium lactis* and *Streptococcus thermophilus*. Diarrhoea can occur when the absorption of sodium in the intestinal villi is impaired while the secretion of chloride in the intestinal crypts secretion continues or is increased, causing abnormal secretion of water and salts into the small intestine. Laboratory experiments with rat distal colon cells showed that SCFAs stimulated the absorption of sodium. Thus, the underlying mechanism proposed for the anti-diarrhoeal effect of probiotics is that they can increase production of short-chain fatty acids in the colon which can then stimulate absorption of sodium, leading to normal fluid secretion in the intestine. Another beneficial effect of probiotics is a decrease in intestinal permeability and invasion by pathogenic microorganisms. Other beneficial effects of probiotics include improved intestinal health, improved symptoms of lactose intolerance, and reduction in risk of various diseases.

Key points on clinical benefits of probiotic consumption:

- Duration and frequency of diarrhea are reduced
- Intestinal permeability and invasion by pathogenic microorganisms are reduced
- Intestinal health and symptoms of lactose intolerance are improved
- Risk of various diseases is reduced

6.2.3 *Effects of probiotics on the immune system*

The gut mucosa is a major site for lymphocytes to contact with antigens as large amounts of antigens enter the gut daily. The microbiota plays a key role in preventing the attachment, growth, and invasion of pathogenic microorganisms on the surface of the intestine. The ingested probiotics can transiently reside in the lower part of the gastrointestinal tract where they interact with epithelial cells and other immune cells, including dendritic cells and M cells in the intestine where they can modulate both innate and adaptive immune responses.

The most obvious effect of probiotics is to decrease adherence and invasion of pathogens by preventing them from binding and penetrating the intestinal epithelial cells. Probiotics enhance mucosal integrity through blocking apoptosis and promoting the survival of intestinal epithelial cells. Probiotics can also inhibit growth of pathogens by producing antibacterial substances such as bacteriocins and acids (lactic, acetic and propionic acids). Probiotics play a role in modulating adaptive immunity by stimulating the production of antibodies and T lymphocytes, and modifying cytokine expression. It should be noted that the effects of probiotics on the immune system is dependent not only on the specific strain but also on the dose, route, and frequency of administration. *In vitro* studies have shown that certain *Lactobacilli* strains are able to induce pro-inflammatory cytokines such as IL-1, tumor necrosis factor (TNF)-α, IL-12, and IFN-γ (Figure 6.2). IL-12 favours development of Th1 cells which can enhance the activity of natural killer (NK) cells, thus promoting cell-mediated immunity. Probiotics can also decrease production of IL-6 by macrophages at mucosae and increase production of anti-inflammatory cytokine IL-10 (Figure 6.2). Other strains have been shown to promote anti-inflammatory cytokines such as TGF-β and thymic stromal lymphopoietin (TSL). Mice treated with probiotics have been shown to induce a reduction in the levels of serum pro-inflammatory cytokines and serum IgA and IgM. Secretory IgA and production of the anti-inflammatory cytokine IL-10 in the intestine are increased. Differentiation of immature dendritic cells to regulatory dendritic cells are promoted by these anti-inflammatory cytokines. The development of T regulatory (Treg) cells is induced by the regulatory dendritic cells. This allows the Tregs to exert anti-inflammatory effects through inhibitory actions on Th1 and Th2 cells and likely also on Th17 cells. Th17 cells are relatively abundant at the intestinal mucosae and are involved in worsening of autoimmunity. Thus, *in vitro* studies with probiotics provide evidence that inflammatory responses are suppressed and anti-inflammatory cells such as Tregs are recruited, resulting in maintenance of homeostasis at the intestinal mucosae (Figure 6.2).

Several human clinical trials have shown that regular consumption of probiotics increases phagocytosis by polymorphonuclear leukocytes and monocytes, bactericidal activity and NK cell activity. It is possible that probiotics enhance production of IL-12 from macrophages resulting in differentiation of naïve T cells to Th1 cells, which augments activity of NK cells. *Lactobacilli* can differently modulate macrophages and DCs and the effects are strain-specific. Thus, probiotics can potentially be useful for the treatment of chronic inflammatory diseases.

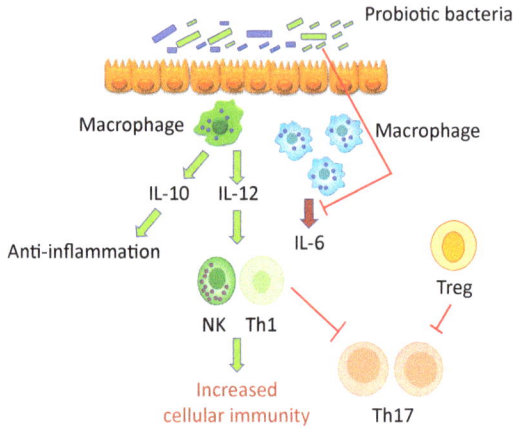

Figure 6.2   Effects of probiotics on macrophages, T cells, NK cells and dendritic cells.

Key points on probiotics:

- Inhibit growth of pathogens by producing antibacterial substances
- Decrease adhesion of pathogens to intestinal epithelial cells
- Block binding of pathogens to intestinal endothelial cells
- Enhance mucosal integrity
- Increase secretory IgA in intestine
- Increase phagocytosis
- Increase bactericidal activity
- Reduce production of pro-inflammatory cytokines (depending on the strain)
- Increase produce of IL-10, TGF-β and TSL
- Enhance IL-12 production and promote differentiation of naïve T cells to Th1
- Increase NK cell activity
- Promote maturation of dendritic cells
- Induce Tregs
- Tregs and Th1 inhibit Th17 activity

Clinical trials with probiotics and prebiotics for treating inflammatory conditions including inflammatory bowel diseases, diarrhoeal diseases, gastrointestinal infections and allergy have been hampered by a lack of convincing clinical trials with reproducible results. The health benefits of probiotics need more clinical research. Although beneficial effects of probiotics have been reported in some cases, there has been concern about the risk of infection and sepsis in immunocompromised patients. Further studies are also required to determine whether different species and strains of the microorganism in probiotics may have different effects on the immune system. Other factors such as the dose, route, and frequency of probiotic administration will also need to be considered in these clinical trials.

6.3 Prebiotics

6.3.1 *Introduction*

Prebiotics are indigestible fibres that selectively stimulate the growth and/ or activity of beneficial gut microbiota. Prebiotics have a beneficial effect on host health. Examples are inulin, galacto-oligosaccharides, and fiber carbohydrates (including β-glucans, lignin, cellulose, pectin and gums). These oligosaccharides are present in foods such as cereals, soybeans, beans, onions and artichokes (Figure 6.3). These fibres are not digestible in the upper gastrointestinal tract but once they get to the colon, they stimulate the activity of lactic acid bacteria of the *Bifidobacterium* and *Lactobacillus* genera which produce SCFAs, and particularly acetate, pro-pionate, butyrate, and lactate.

6.3.2 *Effects of prebiotics on the immune system*

Prebiotics have positive effects on the intestinal barrier by preventing gut pathogens from adhering to the gut epithelium. They also have an impact on the gut immune system through production of SCFAs by gut microbi-ota. One of these SCFAs, butyrate, has been the most studied. Butyrate

Figure 6.3   Examples of dietary sources of prebiotics.

plays a major role in providing the major energy source for colonic epithelial cell proliferation. Butyrate also acts as a barrier to prevent colonisation of pathogens at the gut epithelium. In addition, butyrate has the ability to reduce oxidative DNA damage. *In vitro* studies have shown that butyrate enhanced the intestinal barrier by regulating the assembly of tight junction proteins. Other beneficial effects of prebiotics include prevention of diarrhoea or constipation, enhancement of lipid metabolism, and increased mineral adsorption such as calcium absorption from the gut, thus favouring increases in bone mineral density.

Animal studies have shown that high fibre consumption promote or increase growth of gut *Bifidobacterium* leading to an increase in the concentrations of SCFAs. These SCFAs such as butyrate, propionate and acetate play key roles in the regulation of expression of immune system genes. Butyrate and propionate exhibit strong anti-inflammatory properties by inhibiting pro-inflammatory cytokines such as TNF-α and IL-8 production from immune and intestinal epithelial cells. Butyrate has been shown to inhibit maturation of dendritic cells, T cell priming by dendritic cells and pro-inflammatory cytokine production resulting in reduced Th17 activity (Figure 6.4) and dampening of inflammation. The benefits of butyrate and propionate on intestinal health is as evidenced by the reduction of colitis and other inflammatory bowel disorders such as

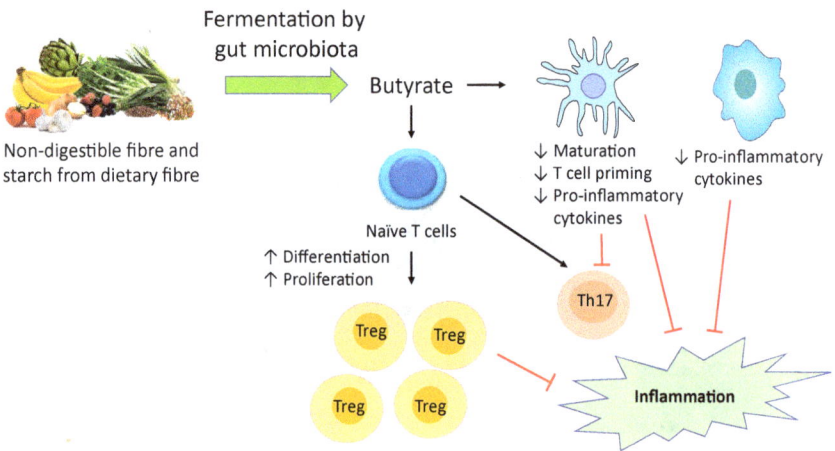

Figure 6.4   Effects of butyrate on immune cells.

Crohn's disease. In an experimental model of inflammatory bowel disease, butyrate has been shown to increase numbers of Treg cells and reduce production of IFN-γ, which can lead to reduced inflammation (Figure 6.4).

Animal studies have shown that consumption of high fibre content can increase the proportion of $CD8^+$ and $CD4^+$ T cells, as well as NK cell activity in the lamina propria and peripheral blood. Animal studies have also demonstrated that prebiotics increased phagocytic activity and enhanced IgA secretion.

Thus, animal studies have shown that prebiotics can modulate the immune system by enhancing innate and adaptive immunity and dampening inflammation. The modulation of inflammation will help to improve inflammatory intestinal disorders such as irritable bowel syndrome.

6.4 Synbiotics

Synbiotics are referred to as products that contain a combination of both probiotics and prebiotics. The usual combinations are fructo-oligosaccharides with *Bifidobacteria* or lactitol with *Lactobacilli*. Reduction in inflammatory cytokines such as TNF and IL-1β and dramatic effects on mucosal regeneration of the gut epithelium were found in patients administered with synbiotics for one month.

Experimental evidence shows that probiotic and symbiotic intake can increase intestinal levels of beneficial commensal microorganisms *Lactobacilli* and *Bifidobacteria*, with reduction in *Enterobacter*. The composition of intestinal microbiota has been reported to change with aging with a decrease in the number of beneficial commensal microorganisms such as *Lactobacilli* and *Bifidobacteria*, and an increase in numbers of facultative anaerobes and Gram-negative bacteria (mainly *Enterobacter*). Thus, age-related changes in intestinal microbiota can be counteracted by probiotics and synbiotics and therefore the risk of infections can be reduced.

Summary of Chapter 6

This chapter explains the effects of probiotics and prebiotics on the immune system. Common examples of probiotics belong to the *Lactobacilli* and *Bifidobacteria* species and they are consumed in yogurt and other fermented foods. Probiotics have been shown to promote innate and adaptive immune responses of the host. Probiotics can increase the barrier function of gut epithelium and stimulate the production of antibodies, T lymphocytes and cytokines. The effects of probiotics on the immune system is dependent not only on the specific strain but also on the dose, route, and frequency of administration. *In vitro* studies have shown that certain *Lactobacilli* strains are able to induce pro-inflammatory cytokines such as TNF-α, IL-1, IL-6, IL-12, and IFN-γ while other strains promote anti-inflammatory cytokines such as TGF-β and TSL. These anti-inflammatory cytokines promote the differentiation of immature dendritic cells to regulatory dendritic cells, which induces development of Tregs. Tregs exert anti-inflammatory effects through inhibitory actions on Th1 and Th2 cells and likely also on Th17 cells. Th17 cells are relatively abundant at the intestinal mucosae and are involved in worsening of autoimmunity. Probiotics can also decrease production of IL-6 by macrophages at mucosae and increase IL-10 production. Mice treated with probiotics have been shown to induce a reduction in the levels of serum pro-inflammatory cytokines, IgA, and IgM, and elevate the levels of anti-inflammatory cytokine IL-10 and secretory IgA in the intestine. Thus, *in vitro* studies provide evidence that probiotics suppress inflammatory responses and recruit anti-inflammatory cells such as Tregs, resulting in maintenance of homeostasis at the intestinal mucosae.

Probiotics may have clinical relevance in inflammatory bowel disorders, diarrhoeal diseases, gastrointestinal and allergic diseases. However, more clinical research is needed. Prebiotics are oligosaccharides present in foods such as cereals, soybeans, beans, onions and artichokes which are not digestible in the upper gastrointestinal tract but once they get to the colon, they are selectively fermented by resident bacteria into SCFAs, and particularly acetate, propionate, butyrate, and lactate. The beneficial effects of prebiotics include inhibition of adherence of invading pathogens to the gut epithelium, prevention of diarrhoea or constipation, enhanced lipid metabolism, and increased mineral absorption such as calcium absorption from the gut, thus favouring increases in bone mineral density. Butyrate has positive effects on the barrier function, proliferation of colonic epithelium and reducing oxidative DNA damage. Butyrate can inhibit maturation of dendritic cells, T cell priming and pro-inflammatory cytokine production by dendritic cells, resulting in reduced Th17 activity and reduced inflammation. Butyrate can increase differentiation of naïve T cells to Tregs and also increase proliferation of Tregs, leading to reduced inflammation. In an experimental model of inflammatory bowel disease, consumption of high fibre diet has been shown to increase the number of Tregs and reduce production of IFN-γ, leading to a down-regulation of inflammation. In addition, prebiotics can increase phagocytic activity and enhance IgA secretion. Synbiotics are referred to as products that contain a combination of both probiotics and prebiotics, usually fructo-oligosaccharides with *Bifidobacteria* or lactitol with *Lactobacilli*. They have been shown to reduce inflammatory cytokines such as TNF and IL-1 and have the ability to regenerate mucosa of the gut epithelium in patients administered with synbiotics for one month. Probiotics and synbiotics may reduce risk of infections in the elderly as they counter the age-related changes in intestinal microbiota.

Chapter 7

Autoimmunity and nutrition

Learning objectives

At the end of studying this chapter, you should be able to:

1. Explain what autoimmunity is
2. Explain the immunological basis of tissue damage in autoimmune diseases
3. Explain the potential role of micronutrients in preventing and treating autoimmune diseases such as Type I diabetes, rheumatoid arthritis, systemic lupus erythrematosus and autoimmune thyroid diseases (Hashimoto's thyroiditis and Grave's disease)

7.1 Introduction

Autoimmunity refers to immune responses to self-antigens due to loss of tolerance. In autoimmune diseases, the immune system is activated by self-antigens and pathological damage leads to diseased states.

Autoimmune diseases can be divided into two types, namely, organ-specific and systemic autoimmune diseases. In organ-specific autoimmune disease ssuch as type I diabetes mellitus, immune destruction of β-islet pancreatic cells results in lack of insulin production and elevated blood glucose levels. Rheumatoid arthritis is characterised by swelling, pain and functional impairment of the joints. Systemic lupus erythematosus (SLE) is a systemic autoimmune disease characterised by various autoantibodies with several organ complications such as skin rashes, cardiovascular disease and glomerulonephritis. The most common thyroid autoimmune diseases are Hashimoto's thyroiditis and Graves' disease, whereby the thyroid gland produces abnormal levels of thyroid hormones.

7.2 Immunological basis of self-tolerance

The mechanisms that contribute to self-tolerance to prevent healthy cells and tissues of the body from mounting an immune attack resulting in an autoimmune disease are: (1) negative selection of T and B cells in the thymus and bone marrow, respectively. Cells that react with autoantigens are eliminated at these sites; (2) autoreactive T and B cells that reach the peripheral circulation are controlled due to induction of anergy; and (3) autoimmune responses are suppressed by T regulatory cells (Tregs).

There are at least two types of Tregs, namely, naturally occurring Tregs in the thymus which are responsible for clonal deletion of self-antigen specific T cells and inducible Tregs which are in the periphery and are inducible from naïve T cells in response to cytokines. Defects in Treg function leads to upregulation of immune response to self-antigen. TGF-β is needed to induce expression of Foxp3 in Tregs. Together with interleukin (IL)-2, TGF-β can induce naïve T cells to differentiate to Foxp3$^+$ Tregs.

7.3 Immunological basis for loss of self-tolerance

In recent years, studies in animal models have shown that an imbalance in Th17/Tregs contributes to altered homeostasis, resulting in autoimmune diseases. TGF-β and IL-6 induce differentiation of naïve T cells to Th17 cells, which are regarded as cells that are pro-inflammatory. The relationship between Th17 cells and Tregs must remain balanced to preserve functional immunity and health of the body (Figure 7.1). Increased Th17 cells or reduced Tregs can induce autoimmunity.

7.4 Influence of nutrition on the onset and progression of autoimmune diseases

The relationship between micronutrients and Th17/Treg balance will be elaborated in the sections specific to autoimmune diseases. We will first

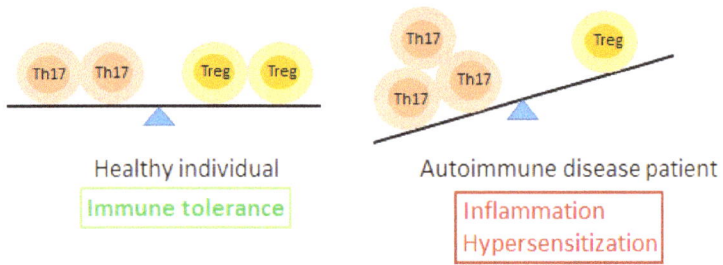

Figure 7.1 Imbalance of Th17/Tregs induces autoimmunity.

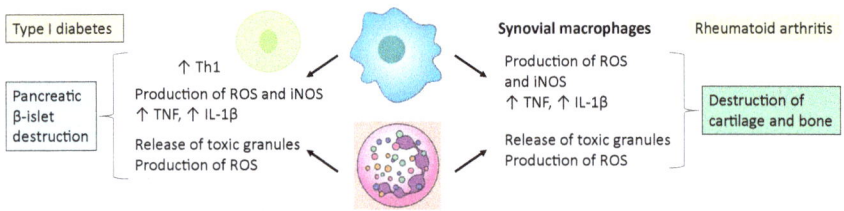

Figure 7.2 Mechanism of tissue damage by macrophages and neutrophils in type I diabetes and rheumatoid arthritis.

address the issue of chronic inflammation in autoimmunity. In general, autoimmune diseases are chronic inflammatory diseases involving oxidative stress and progression to tissue destruction. Macrophages and neutrophils release reactive oxygen species and toxic granules to destroy pancreatic islet cells in type I diabetes and cartilage and bone in rheumatoid arthritis (Figure 7.2).

Thus, in theory, foods with anti-inflammatory constituents are highly recommended and those that promote inflammation should be limited. Overall, a well-balanced diet that includes plenty of fruits, vegetables, and whole grains and moderate amounts of meats, poultry, and oily fish is recommended. Foods with possible anti-inflammatory properties include fruits and vegetables and they are also rich in antioxidants. In addition, foods containing omega-3 fatty acids, such as fish, nuts, flaxseed, canola oil, and olive oil may also help fight inflammation. On the other hand, saturated fats should be limited as they contribute to inflammation. These include animal fat, processed meat products, high fat dairy foods and fried foods.

The influence of micronutrients in foods in specific autoimmune diseases will next be explained.

7.5 Type I diabetes mellitus

Majority of cases of diabetes in childhood is Type I diabetes mellitus. Patients have abnormal increase in blood glucose levels due to lack of insulin production as a result of immune destruction of β-islet pancreatic cells. Destruction of pancreatic islet cells is caused by islet-specific T cells and cytokines such as tumour necrosis factor (TNF) produced, which promote apoptosis of β-islet pancreatic cells. Patients with type I diabetes mellitus have increased production of several autoantibodies.

Nutritional factors associated with type I diabetes mellitus include the short duration of breast feeding and exposure to cow's milk before 4 months of age that were shown with animal models. However, this is controversial as most human studies have focused on early exposure to proteins from cow's milk. Other dietary linkages relate to the higher consumption of food containing nitrosamines, which is known to be present in processed meat, by pregnant mothers.

In recent years, there has been increasing interest in the role of vitamin D in the pathogenesis and prevention of type I diabetes. Type I diabetes is characterised by autoimmune destruction of insulin-producing β-cells in the pancreas. In a non-obese type I diabetes mouse model, vitamin D_3 ($1,25(OH)_2D_3$) was shown to reduce the incidence of diabetes. In human studies, children who received vitamin D supplementation had a 33% reduction in the risk of developing type I diabetes as compared with non-supplemented children. This may be due to the immunosuppressive role of vitamin D on antigen presenting cells (macrophages and dendritic cells) and activated T lymphocytes (Figure 7.3). Antigen

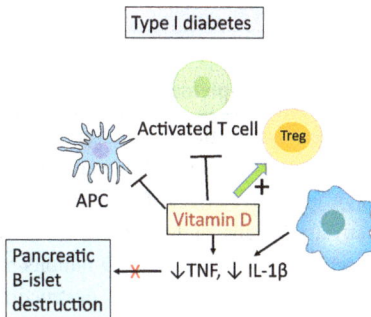

Figure 7.3   Role of vitamin D in preventing pancreatic β-islet destruction in Type I diabetes mellitus.

presentation and expression of co-stimulatory molecule by dendritic cells and secretion of pro-inflammatory cytokines such as TNF-α are downregulated. Vitamin D promotes the induction of Tregs (Figure 7.3). This is consistent with data from animal studies showing inhibition of pro-inflammatory cytokine (IL-1β and IFN-γ) synthesis by vitamin D prevents β-islet pancreatic cell destruction.

7.6 Rheumatoid arthritis

Rheumatoid arthritis is characterised by swelling, pain and functional impairment of the joints. It is usually presented with signs of systemic inflammation as shown by the increase in plasma levels of acute phase proteins and cytokines. Synovial fluid from these patients contain high levels of pro-inflammatory cytokines such as TNF and IL-1. This results in recruitment of inflammatory cells to the joint which destroy the bone and cartilage there, causing the characteristic joint bone deformity and pain.

The effects of micronutrient supplementation in rheumatoid arthritis are summarised in Table 7.1. Consumption of oily fish and fish oils containing EPA and DHA has shown clinical improvement in patients with rheumatoid arthritis such as reduced joint rigidity and pain. Consumption of these fish oils results in incorporation of fatty acids into the cell membrane and consequently alter the fatty acid composition. These fatty acids can decrease antigen presentation, T cell proliferation and consequent production of cytokines.

Table 7.1   Summary of micronutrient suppplementation in some autoimmune diseases.

Autoimmune Disease	Micronutrient Supplementation and Potential Mode of Action
Type I diabetes	Vitamin D supplementation in children: 33% reduction in risk of developing type I diabetes. This may be due to: (1) immunosuppressive effect of vitamin D on antigen presenting cells and activated T cells, and (2) induction of Tregs. Animal studies show that vitamin D inhibits pro-inflammatory cytokine production and prevents destruction of pancreatic β-islet cells.

Table 7.1   (*Continued*)

Autoimmune Disease	Micronutrient Supplementation and Potential Mode of Action
Rheumatoid arthritis	Consumption of oily fish and fish oils containing EPA and DHA has shown some clinical benefits such as reduced joint rigidity and pain. This could be due to alteration in cell membrane composition, decreased antigen presentation, T cell proliferation and production of cytokines. Vitamin A can induce production of Tregs which can suppress immune activation, inhibit IL-17 production and control inflammation. IL-17 drives production of autoantibodies which is suppressed by Tregs.
SLE	Animal studies have shown ω-3 PUFAs suppress macrophage activity and production of cyclooxygenase metabolites. Human studies show that supplementation of fish oils reduces proteinuria and improves glomerulonephritis, thus delaying the progression of renal disease. Intake of flax seeds by patients with lupus nephritis reduces serum creatinine and proteinuria. In animal models, it has been shown that retinoic acid (vitamin A metabolite) inhibits formation of Th17 cells and promotes production of Tregs. For SLE patients, intake of retinoic acid reduces proteinuria, antibody levels of dsDNA and complement. In animal models, vitamin D_3 improved survival, bone health and immunity, and reduced proteinuria. Supplementation with vitamins B_6, B_{12} and folate appears to be beneficial for patients. Reduction of plasma homocysteine by these vitamins can reduce severity of atherosclerosis, which is one of the complications of SLE. Studies in a mouse model show that vitamin C reduces IgG and antibodies against dsDNA. In patients, vitamin C supplementation has been shown to reduce risk of inflammatory activity and prevent cardiovascular complications. Zinc restriction in diet of mice reduced lymphoproliferation, dsDNA antibodies, and improved glomerulonephritis. Animals fed with selenium-rich diet showed reduced inflammation, reduced dsDNA antibodies, and prolonged survival. Adequate consumption of calcium and vitamin D_3 is recommended for SLE patients as there is risk of glucocorticoid-induced osteoporosis and fractures.

(*Continued*)

Table 7.1   (*Continued*)

Autoimmune Disease	Micronutrient Supplementation and Potential Mode of Action
Hasimoto's thyroiditis (autoimmune thyroid disease)	Deficiency in vitamins C and E can cause oxidative stress and lead to damage of thyroid gland. Deficiencies in vitamin B_{12}, magnesium, and zinc interfere with production of TSH. Deficiencies in vitamins B_2 and C are associated with reduced iodine uptake, which then leads to reduction of thyroid hormone synthesis. Iron deficiency is found in 60% of patients with hypothyroidism. Iron is a cofactor in iodine peroxidase, which converts iodide to iodine in the thyroid gland. The production of thyroid hormones T3 and T4 requires iodine and tyrosine. Thus, lack of thyroid peroxidase activity due to iron deficiency results in impairment of synthesis of T4. Conversion of T4 to T3 requires selenium. Deficiency of selenium increase risk of autoimmune disease development. Selenium enhances Treg cell activity and suppresses cytokine secretion, thus follicular cell destruction and thyroiditis are prevented. Excess iodine increases frequency of autoimmune thyroiditis. There is increase in Th17, inhibition of Tregs and increased TRAIL resulting in destruction of thyroid follicular cells. Patients with Hashimoto's thyroiditis have been found to have low blood levels of vitamin D, and supplementation was associated with an important decrease in TPOAb, TgAb and TSH levels. Decreased parathyroid activity disrupts production of vitamin D, resulting in dysfunction in intestinal absorption of calcium. Supplementation of magnesium in magnesium-deficient Hashimoto patients have been shown to decrease serum levels of thyroglobulin.
Grave's disease	Vitamin D and selenium supplementation is recommended for patients with these deficiencies.

7.7 Vitamin A contributes to balance of Th17/Tregs and downregulation of antibody production

Vitamin A plays a central role in maintaining and restoring the integrity and function of all mucosal surfaces in addition to modulating innate and adaptive immunity. TGF-β and retinoic acid (vitamin A metabolite) produced by the intestinal epithelium play a key role in the differentiation of peripheral naïve T cells to Tregs (Figure 7.4). Tregs are a subset of

Vitamin A

- induces production of T regulatory cells which can suppress immune activation
- inhibits IL-17 production and controls inflammation

IL-17 drives production of autoantibodies which is suppressed by T regulatory cells.

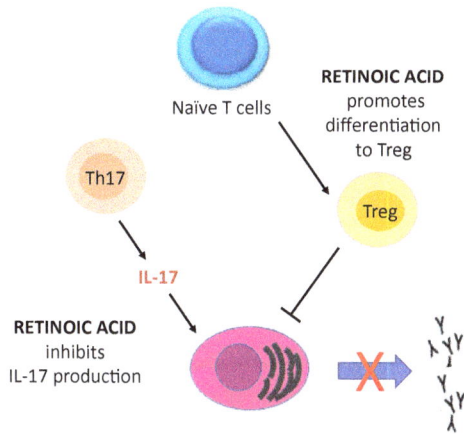

Naïve T cells

RETINOIC ACID promotes differentiation to Treg

Th17

Treg

IL-17

RETINOIC ACID inhibits IL-17 production

Figure 7.4   Role of retinoic acid (vitamin A metabolite) in downregulating Th17 cells (pro-inflammatory), upregulating Tregs (anti-inflammatory) and suppressing production of autoantibodies.

T cells that can suppress immune activation and thus modulate intestinal inflammation. Tregs can control excess of IL-17, which is a cytokine that supports the inflammatory cascade. IL-17 has been shown to drive production of autoantibodies in rheumatoid arthritis. Inhibition of IL-17 production by Th17 cells by vitamin A may contribute to the balance of Th17/Tregs and maintain homeostasis. Thus, adequate consumption of vitamin A is important for immunity as well as to direct immune tolerance to self-antigens or non-harmful antigens.

7.8 Systemic Lupus Erythematosus (SLE)

7.8.1 *Introduction*

SLE is a multi-organ systemic autoimmune disease characterised by several autoantibodies such as elevated levels of anti-double stranded (ds) DNA and low levels of complement. Complications of SLE include skin rashes, cardiovascular disease and glomerulonephritis. These complications can be indicated by high levels of plasma homocysteine and increased proteinuria. Elevated levels of TNF and other pro-inflammatory cytokines are found in SLE patients. Thus, reduction of inflammation appears to be a rational nutritional strategy for SLE patients. The effects of micronutrient supplementation in SLE are summarised in Table 7.1.

7.8.2 *SLE and ω-3 Polyunsaturated Fatty Acids (PUFAs)*

ω-6 PUFAs are the precursors of pro-inflammatory eicosanoids such as prostaglandins (PGD_2, PGE_2, and $PGF_{2\alpha}$), thromboxane (TXA_2) and leukotrienes (LTA_4) which can influence or increase risk for other inflammatory conditions in SLE patients. In contrast, ω-3 PUFAs are precursors of EPA and DHA which generate resolvers of inflammation, resolvins and protectins. Fish oil is a rich source of ω-3 PUFAs.

The anti-inflammatory effects of ω-3 PUFA such as EPA and DHA which are precursors of resolvins and protectins have been established *in vitro* and in animal studies. The activity of macrophages and the production of cyclooxygenase metabolites were suppressed.

SLE patients frequently have increased blood triglyceride levels, low high-density lipoprotein and are at increased risk for cardiovascular events. A study showed that consumption of a diet low in saturated fat and cholesterol by SLE patients led to a significant reduction in levels of low-density lipoprotein and cholesterol. Supplementation with fish oil reduced proteinuria and improved glomerulonephritis. The reduction in inflammation via ω-3 PUFA supplementation has been proposed to delay progression of renal diseases in several clinical studies. Other studies have shown that dietary intake of 30–45 grams per day of flax seed (a rich source of the essential fatty acid, α-linolenic acid, which is a precursor of ω-3 PUFAs) by patients with lupus nephritis, reduced serum creatinine and proteinuria. Similarly, severity of symptoms in SLE and triglyceride levels was reduced when the diet was supplemented with fish oils.

7.8.3 *Role of vitamin A in SLE*

In animal models of SLE, the vitamin A metabolite, retinoic acid, inhibited the development of pro-inflammatory Th17 cells and promoted the development of anti-inflammatory Tregs. Treatment of SLE patients with retinoic acid appeared promising as there was reduced proteinuria, reduced levels of anti-ds DNA and improved levels of complement.

7.8.4 *Vitamin D*

MRL/lpr mice spontaneously develop autoimmune lupus syndrome. When these mice received vitamin D_3 supplementation, proteinuria was reduced and bone health and survival were improved. However, it is still unclear whether vitamin D supplementation is beneficial for SLE patients with low serum levels of $1,25(OH)_2D_3$.

7.8.5 *Vitamin E*

It is also unclear whether vitamin E supplementation in SLE patients is useful. In the mouse model of lupus, the appearance of autoimmunity is delayed and survival is increased. This may be due to modulation of levels of inflammatory cytokines by vitamin E.

7.8.6 *Vitamin B complex*

One of the complications of SLE is atherosclerosis. In this condition, high plasma homocysteine levels are found. To reduce the accumulation of plasma homocysteine in plasma, a higher consumption of vitamins B_6, B_{12} and B_9 (folate) is recommended. These vitamins are required as cofactors for the breakdown of homocysteine (Figure 7.5).

Homocysteine can be converted back to methionine by the addition of a methyl group from betaine (which can be obtained from shellfish, wheat germ, wheat bran and certain plants such as beets and spinach) or methyltetrahydrofolate with the help of vitamin B_{12}. Homocysteine can also be

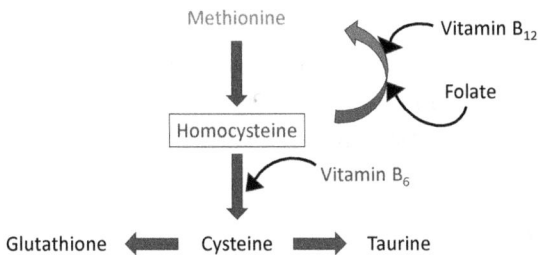

Figure 7.5 Metabolic pathway for glutathione generation from homocysteine and roles of vitamins B_{12}, B_9 (folate) and B_6.

broken down to cysteine in a reaction which is catalysed by vitamin B_6. Cysteine can then be used for the formation of glutathione, an antioxidant, or oxidised to form the amino acid taurine.

In mice models, folate supplementation minimises the symptoms of SLE and increases survival. Supplementation with high doses of folate and vitamin B_6 has been shown to reduce the severity of SLE in a Japanese study.

Thus, supplementation with vitamins B_6, B_{12} and folate appears to have beneficial effects in SLE patients.

7.8.7 *Vitamin C*

The reduction in IgG and anti-dsDNA levels in mice models of SLE has been reported. In SLE patients, consumption of vitamin C has been shown to reduce the risk of SLE inflammatory activity and prevent cardiovascular complications.

7.8.8 *Zinc*

Zinc restriction in the diet for MRL/lpr mice resulted in reduced lymphoproliferation and antibodies against dsDNA, and an improvement in glomerulonephritis. This can also lead to increased serum corticosteroids which can control SLE. However, since zinc deficiency causes immune dysfunction, zinc restriction in the diet is not likely to be of benefit to SLE patients.

7.8.9 *Selenium*

A diet rich in selenium reduces anti-dsDNA autoantibodies, inflammation and prolongs survival in murine models of SLE. It is unclear if selenium supplementation is beneficial to SLE patients.

7.8.10 *Calcium*

Adequate calcium consumption is important for patients with SLE as they are at risk for glucocorticoid-induced osteoporosis and fractures. They are more likely to undergo bone fractures as compared with healthy women of the same age. Supplementation of vitamin D (20 μg or 800 UI) and calcium (>1,500 mg) is recommended for these patients with inadequate dietary intake.

Overall, ω-3 PUFA, vitamin D and minerals such as calcium and selenium can promote a beneficial protective effect against tissue damage and suppression of inflammatory activity for SLE patients. Zinc should be restricted to prevent disease aggravation.

7.9 Autoimmune thyroid diseases and micronutrients

7.9.1 *Introduction*

The thyroid is an important endocrine organ for vital physiological functions including regulation of respiration, heart and nervous system, brain development, body temperature, blood calcium levels, carbohydrate and fat metabolism and cholesterol levels. Hypothyroidism is due to inadequate levels of thyroid hormones and is the main feature of Hashimoto's disease, whereby autoantibodies mount an immune attack on the thyroid gland resulting in malfunction of the thyroid and inadequate production of thyroid hormones. Increased cholesterol levels can induce hypothyroidism. In contrast, hyperthyroidism is due to excessive secretion of thyroid hormones. The main cause of hyperthyroidism is Graves' disease, where agonistic autoantibodies to thyroid stimulating hormone (TSH) bind to the TSH receptor to stimulate production of thyroid hormones. Both diseases are referred to as autoimmune thyroid diseases and generally occur during middle age, and they are more common in women than in men.

7.9.2 *Autoimmune thyroid diseases and micronutrient deficiencies*

Micronutrient deficiencies have been observed in patients with autoimmune thyroid diseases. Some minerals such as iodine, iron, selenium and zinc are required for the synthesis and metabolism of thyroid hormones. Other micronutrient deficiencies include deficiencies in protein, vitamins (A, C, B_1, B_2, B_5, B_6, B_{12}) and other minerals (magnesium, sodium,

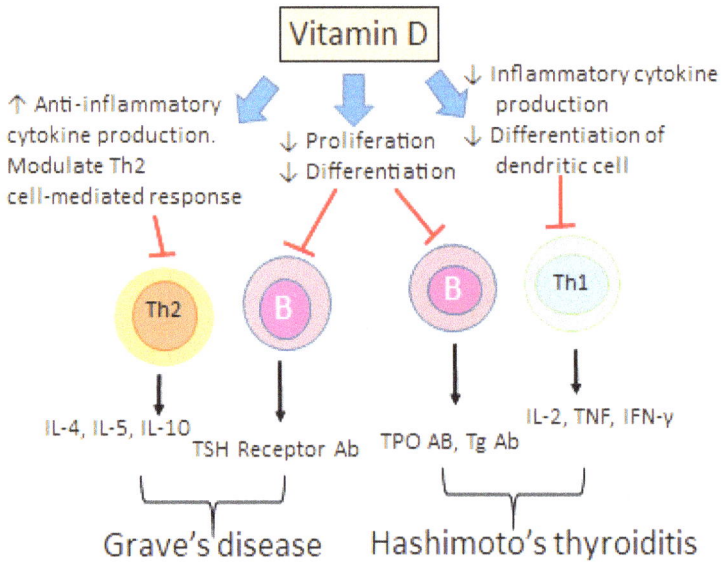

Figure 7.6   Role of vitamin D in downregulating autoreactive T cells and antibodies in autoimmune thyroid diseases.

potassium, phosphorus and chromium). Vitamin D acts to: (1) reduce proliferation and differentiation of B cells into plasma cells, (2) inhibit Th1 cell proliferation and Th1 cytokine (IFN-γ, IL-2, and TNF-α) production, and (3) modulate Th2 cells and increase anti-inflammatory cytokine (IL-10) production (Figure 7.6). Vitamin D supplementation reduces Grave's disease recurrence and has been associated with a decrease in thyroid peroxidase antibodies (TPOAb) and thyroglobulin (TgAb) levels.

7.9.3 *Consequences of micronutrient deficiencies and excess iodine in autoimmune thyroid diseases*

Hashimoto's thyroiditis is often characterised by hypothyroidism, elevated thyroid autoantibodies such as TgAb and TPOAb, with thyroid glandular lymphocytic infiltration. Immune attack on the thyroid gland results in inadequate production of thyroid hormones. The effects of micronutrient supplementation in Hashimoto's thyroiditis are summarised in Table 7.1.

Deficiency in vitamins C and E which are antioxidant vitamins can cause oxidative stress and lead to damage to the thyroid gland by free radicals and membrane lipid peroxidation.

Vitamin B_{12}, magnesium, and zinc are required to make TSH in the pituitary gland. Deficiencies in vitamin B_{12}, magnesium and zinc can interfere with production of TSH. TSH stimulates the production and release of thyroid hormones, namely, thyroxine (T4) and triiodothyronine (T3). The production of T4 and T3 requires iodine and tyrosine. Iodide is transported from the blood into the thyroid gland by the sodium/iodide pump. The sodium/iodide pump activity requires vitamins B_2 and C. Deficiencies in them are associated with reduced iodine uptake by the thyroid gland and lead to reduction in the synthesis and secretion of thyroid hormones.

Sufficient iron intake is necessary for proper thyroid function. Iron is necessary in the production of thyroid hormones, and its deficiency blocks the activity of thyroid peroxidase. Iron is a cofactor in iodine peroxidase, which converts iodide to iodine in the thyroid gland. As a result, a reduction in the synthesis of thyroid hormones is observed, as well as increases in TSH level and gland volume. Anaemia due to iron deficiency may therefore increase the risk of developing thyroid disease.

Decreased parathyroid activity in autoimmune thyroid disease disrupts the production of the active form of vitamin D. This results in abnormal levels of vitamin D in the blood and leads to dysfunction in intestinal absorption of calcium. Blood vitamin D levels should be constantly monitored and vitamin D and/or calcium supplements should be recommended to patients with vitamin D deficiency. Patients with Hashimoto's thyroiditis have been found to have low blood levels of vitamin D. Human intervention studies demonstrated that cholecalciferol (vitamin D_3) supplementation was associated with an important decrease in TPOAb and TgAb levels both in patients with vitamin D sufficiency and deficiency. Another randomised clinical trial showed that supplementation with cholecalciferol resulted in a significant decrease in TgAb and TSH levels.

Deficiency in selenium (blood selenium below 60 µg/L) also increases the risk of autoimmune diseases such as Hashimoto's thyroiditis and cancer, whereas excess selenium (above 140 µg/L) increases risk of diseases such as hyperlipidaemia and type II diabetes mellitus. Normal blood

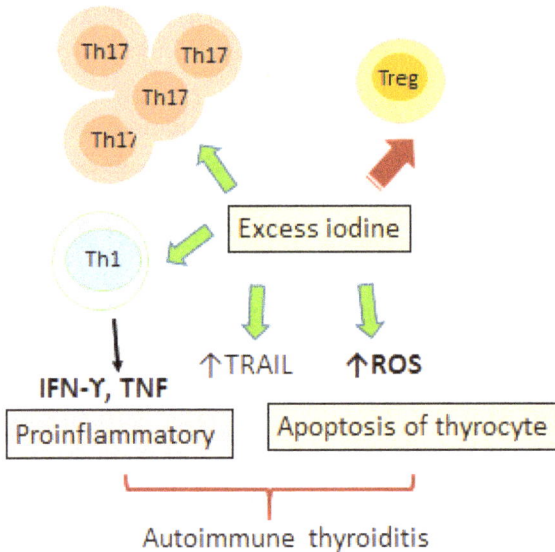

Figure 7.7   Excess iodine stimulates Th1 and Th17 cells (pro-inflammatory), downregulates Tregs and induces apoptosis of thyrocytes resulting in destruction of thyrocytes.

selenium levels of 60–140 µg/L is necessary for health and to inhibit the toxic effect of excessive iodine. Chronic high iodine intake has been associated with increased frequency of autoimmune thyroiditis. Studies in a mouse model show that excess iodine inhibits Treg cell differentiation, increases intra-thyroid infiltrating Th17 cells (pro-inflammatory cells) and increase expression of TNF-related apoptosis-inducing ligand (TRAIL) in thyrocytes, thus inducing apoptosis and tissue destruction (Figure 7.7). Tissue destruction is further enhanced by increased reactive oxygen species generation in the thyrocytes due to excess iodine.

Patients with autoimmune thyroiditis have been found to have low serum selenium levels.

Selenium is a cofactor for many selenoproteins which are involved in antioxidant, redox, and anti-inflammatory processes. Glutathione peroxidase (a selenoenzyme) detoxifies hydrogen peroxide to protect against oxidative damage (Figure 7.8). Conversion of T4 (inactive form) to T3 (active) requires the selenoenzymes iodotyrosine deiodinases (Figure 7.8). Without selenium, no conversion occurs.

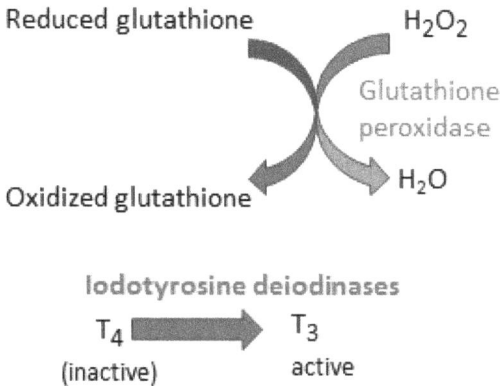

Figure 7.8 Selenium acts as cofactor for glutathione peroxidase and iodotyrosine deiodinases.

In addition, selenium promotes Treg cell activities and suppresses cytokine secretion, resulting in prevention of follicular cell destruction and protection against thyroiditis. Thus, selenium supplementation for patients with autoimmune thyroid diseases and low selenium status can be beneficial.

Lastly, severe magnesium deficiency increases the risk of developing Hashimoto's disease. Supplementation of magnesium in these magnesium-deficient patients has been shown to decrease serum levels of thyroglobulin.

Graves' disease is a common autoimmune disease. These patients have high levels of TSH receptor autoantibodies, hyperthyroidism, goiter, and ophthalmopathy. Results from an interventional study showed that recurrence of Graves' disease occurred earlier in patients not receiving vitamin D supplementation. Selenium supplementation can also be helpful as low selenium levels have been reported in patients with Grave's disease. The effects of micronutrient supplementation are summarised in Table 7.1.

Summary of Chapter 7

Autoimmunity refers to immune responses to self-antigens due to loss of tolerance. Nutritional factors for prevention or treatment of autoimmune diseases is still unclear. Animal studies have shown that vitamin D prevents damage on β-islet pancreatic cells by inhibiting the synthesis of inflammatory cytokines (IL-1β and IFN-γ). This may be due to down-regulation of antigen presentation and expression of co-stimulatory molecule by dendritic cells, thus inhibiting the production of pro-inflammatory cytokines. Vitamin D can induce differentiation of T cells to Tregs which are known for their immunosuppressor functions. In type II diabetes mellitus, vitamin D improves insulin secretion and glucose tolerance. Positive outcomes on vitamin D supplementation in preventing type II diabetes mellitus require further investigation. Consumption of oily fish and fish oils containing EPA and DHA has shown clinical improvement in patients with rheumatoid arthritis such as reduced joint rigidity and pain. Vitamin A may be beneficial due to its ability to induce Tregs and thus downregulate IL-17 production and reduction of autoantibodies in rheumatoid arthritis and SLE. Vitamins D and E, folate and selenium have been shown to improve survival of MRL/lpr mice which spontaneously develop SLE. For SLE patients, supplementation with high doses of folate and vitamin B_6 reduced the severity of SLE. Beneficial protective effect against tissue damage and suppression of inflammatory activity for SLE patients have been shown with ω-3 PUFA, vitamin D and minerals such as calcium and selenium. A delay in renal disease progression has been reported with ω-3 PUFA consumption in SLE patients. Micronutrient deficiencies in patients with autoimmune thyroid diseases include minerals such as iodine, iron, selenium and zinc, and vitamins (A, C, B_1, B_2, B_5, B_6, B_{12}). The impact of these deficiencies on TSH, T4 and T3 synthesis, and thyroid gland damage is explained.

Chapter 8

Nutrition, diet and cancer

Learning objectives

After studying this chapter, you should be able to:

1. Explain key features of cancer cells
2. Explain the link between inflammation and cancer
3. State examples of carcinogens in dietary sources and explain how they can cause carcinogenesis
4. State examples of antioxidants in diet that can reduce cancer risk
5. Provide examples of micronutrients that influence cell growth
6. Describe the immune mechanisms to kill tumours
7. Explain how micronutrients enhance tumour immunity

8.1 Introduction

Cancer is a complex disease. It is widely accepted that cancer arises when cells grow uncontrollably and do not respond to the signals that control cell growth and death. It has been known for a long time that the factors that contribute to the initiation of cancer formation include exposure to carcinogens, chronic inflammation, and poor dietary and lifestyle habits. Carcinogens can cause DNA damage such that the cell undergoes uncontrolled cell cycle progression, proliferation, DNA replication and accumulation of massive mutations in itself and its progeny. In cancer cells, damaged DNA is not repaired and cancer cells do not die. When T cells detect these abnormal cells, they induce apoptosis (programmed cell death) of these abnormal cells. However, the gross changes in the genetic structure of cancer cells often alter their response to death signals. When this occurs, the cancer cells continue to survive in the host. Other factors such as failure of the immune system to eliminate cancer cells and immune evasion by cancer cells allow them to survive and thrive in the host. The current paradigm suggests that the balance of immunosuppressive and effector cells in the tumour microenvironment maintains the tumour cells in a state of immune-mediated dormancy. However, in the equilibrium phase, immunoediting occurs resulting in the emergence of tumour cell variants which escape immune recognition, allowing them to grow progressively.

8.2 Association between infectious agents and cancer

Infection-associated cancers are estimated to contribute to more than 20% of cancer cases worldwide. Examples include hepatitis B or C viruses and liver cancer, human papillomavirus and cervical cancer, and *Helicobacter pylori* and stomach cancer. Other environmental factors, including host nutritional status, can influence infection persistence and the development of dysplasia.

Key points on examples of infectious agents associated with cancer:

- Liver cancer — hepatitis B or C viruses
- Cervical cancer — human papillomavirus
- Gastric cancer — *Helicobacter pylori*

8.3 Relationship between inflammation and cancer

In the context of chronic inflammation induced by infection and cancer risk, the presence of persistent infection and inflammation results in inflammatory responses that can lead to the development of dysplasia and eventual tumour formation. Chronic inflammation recruits infiltrating immune cells and stimulates release of pro-inflammatory cytokines. These immune cells produce excessive reactive oxygen species (ROS) which increases the risk of damage to DNA. Cytokines such as TNF activate the NFκβ pathway, resulting in inhibition of apoptosis and proliferation of cells. Thus, nutrients which can function as antioxidants, or influence the immune system and act to suppress the inflammatory process, can be beneficial for reducing cancer risk.

Key points on how chronic inflammation increases risk of damage to DNA:

- Infiltrating cells produce harmful substances such as ROS
- Excessive ROS leads to oxidative stress which can damage cell structure and DNA
- Pro-inflammatory conditions promote cell proliferation, formation of new blood vessels and inhibit apoptosis

---

8.4 Factors that influence cancer risk

There are several cancer risk factors including family history, tobacco use, nutrition and sedentary lifestyle. Individuals with inherited cancer genes such as *BRCA* genes for breast cancer have an increased risk for breast cancer. A family history of cancer does not guarantee cancer development but means that there is an increased risk and measures should be taken to minimise the risk.

Key points on factors that influence cancer risk:

- Smoke and tobacco use
- Environmental pollution
- Infectious agents
- Obesity
- Hormonal imbalance
- Family history
- Lifestyle

Smoking increases the risk of cancer as tobacco and tobacco smoke are carcinogenic. Industrial pollution, mercury, fine particulates and smoke from factories with unknown contaminants are sources of carcinogens. These chemicals bind to DNA to form DNA adducts resulting in DNA damage. DNA damage is widely accepted as being responsible for cancer initiation and progression.

In addition, cancer is associated with being overweight or obesity, poor nutrition and physical inactivity. Obesity is thought to promote

Figure 8.1 Diet rich in fruits and vegetables reduces cancer risk.

tumourigenesis by raising oestrogen levels and insulin-like growth factors. In obesity, visceral adipose tissues release adipokines and cytokines resulting in the induction of a pro-inflammatory environment.

Diet and lifestyle also influence the risk of developing cancer. Diets high in saturated fat and low in fruits and vegetables can increase risk of cancers such as colon, oesophageal, prostate and breast cancers. Diets high in protein have also been shown to increase cancer risk. Consumption of alcohol and smoked, cured and charbroiled meats, and a sedentary lifestyle can also increase cancer risk. Eating a diet rich in fruits and vegetables is encouraged to minimise cancer risk (Figure 8.1). In June 2020, the American Cancer Society published a guideline on diet and physical activity for cancer prevention.

8.5 Carcinogens in the diet

The process of carcinogenesis is influenced by nutrients and environmental factors. Pro-carcinogens in the diet include aflatoxins which is present in moulds of contaminated peanuts and flour, heterocyclic amines and polycyclic aromatic hydrocarbons that result from charring of meats, acrylamide formed from starchy foods cooked at high temperatures, nitrosamines used or produced in curing of meats, and naturally occurring chemicals in plants and chemicals used in agricultural practices and food handling. These carcinogens can be categorised as exogenous or endogenous carcinogens. Exogenous carcinogens including aflatoxin, contaminating pesticides, food additives and artificial sweeteners such as cyclamates and saccharin have been implicated in bladder cancer. Endogenously synthesized carcinogens including nitrosamines and nitrosamides have been associated with gastric cancer. Other examples of endogenously synthesized carcinogens are nitrites, amines and amides which can derived from digested proteins. They are found in foods such as preserved meats like bacon and sausage. When the meat is cooked, nitrates and nitrites react with amines in the proteins of the food to cause nitrosamines which are carcinogenic. Nitrates are also present in vegetables. They are reduced to nitrites and nitric oxide by bacterial flora in the gut. Lastly, acrylamides produced when certain foods are cooked at very high temperatures can also be a source of carcinogens in our diet.

Key points on diet and cancer:

- Food and environmental factors play an important role in tumour initiation, promotion and progression of cancer

Sources of carcinogens

- Aflatoxin (e.g. mould from peanuts, flour)
- Benzopyrene (e.g. heterocyclic amines) from charbroiled meats
- Acrylamide formed from starchy food cooked at high temperatures
- Nitrosamine from cured meats (preserved or flavoured)
- Naturally occurring chemicals in plants
- Chemicals added in agricultural practices and food handling

Two aspects of diet that can cause carcinogenesis

- Content of exogenous carcinogens
- Endogenous synthesis of carcinogens

Exogenous carcinogens

- Aflatoxin — associated with hepatocellular carcinoma; causes p53 mutation in tumour cells
- Contaminating pesticides
- Some food additives and artificial sweeteners — e.g. cyclamates and saccharin implicated in bladder cancer

Endogenous synthesis of carcinogens

- Nitrites, amines and amides derived from digested proteins
- Nitrates present in vegetables are reduced by bacterial flora in the gut
- Acrylamides released when certain foods are cooked at high temperatures

8.6 Role of folate (vitamin B₉), vitamin D and vitamin A in cell proliferation

The cell cycle consists of four stages or phases, namely, cell division or mitotic phase (M phase), G_1 phase (gap after M phase), S phase (DNA synthesis phase) and G_2 phase (gap after S phase). Control of the cell cycle is important for governing whether a cell should commit to DNA synthesis and proliferation versus growth arrest, DNA repair or apoptosis. Folic acid is required for DNA replication and deficiency in folic acid can reduce cell proliferation by decreasing DNA synthesis. When there is DNA damage in a normal cell, cell cycle is arrested and the cell undergoes DNA repair. This process ensures that errors are corrected and if not, the cell undergoes apoptosis. However, the pathways regulating the cell cycle are frequently dysregulated in cancer. Cancer cells undergo uncontrolled proliferation and continuously proceed through the cell cycle. Micronutrients such as folic acid, vitamin A and vitamin D have been shown to be able to control cell cycle progression. *In vitro* studies show that vitamin A can arrest cells in the G_1 phase of the cell cycle, stopping the cells to proceed to mitosis and DNA synthesis.

8.7 Antioxidants and minimising cancer risk

Consumption of foods high in antioxidants especially fruits, vegetables and whole grains is linked to decreased risk of cancer. Nutrients with antioxidant properties include vitamin C, β-carotene, vitamin E and selenium. Vitamin C functions as an antioxidant and is important in immune function. In the stomach, vitamin C reduces the formation of nitrosamines found in foods such as processed meats. β-carotene stabilises free radicals and prevents damage to cells. Our body converts β-carotene to an active form of vitamin A, retinol. Vitamin E acts as an antioxidant and is a constituent of cell membrane of immune cells. Immune cells are highly susceptible to oxidative damage. The proposed mechanism in cancer prevention is as follows:

— Antioxidants enhance immune function resulting in destruction and elimination of cancer cells

— Growth of cancer cells is inhibited
— Oxidative damage to cell DNA is prevented

Key points on vitamins and minerals as antioxidants or anti-carcinogens:

- Antioxidant properties of vitamins C and E, β-carotene and selenium have anti-carcinogenic effects
- In the stomach, vitamin C reduces the formation of nitrosamines
- Retinoic acid can reduce cancer risk by promoting epithelial cell differentiation and this is believed to reverse squamous metaplasia

---

Key points on dietary influences on cancer development:

Positive influences of diet

- Inhibit carcinogens via antioxidant effects
  — vitamins A, C, E; cruciferous vegetables; soy; green tea
- Block carcinogens from accessing DNA target
  — plant phenols; coumarins; green tea
- Inhibit promotion and progression
  — retinols, protease inhibitors, carotenoids

Negative influences of diet

- Vitamin C (antioxidant) renders cancer treatments less effective

---

Research over the past 20 years has shown that antioxidant supplementation in a healthy individual does not reduce cancer risk. Clinical trials with vitamin C, β-carotene and vitamin E in cancer prevention have been largely disappointing. In a Finnish study, α-tocopherol (vitamin E), β-carotene, a combination of the two or placebo were administered for 5–8 years in 29,133 active male smokers. A higher incidence of lung cancer was found in those receiving β-carotene with a mortality of 8% compared to placebo. In another group receiving α-tocopherol, a lower incidence of prostate cancer versus placebo was reported but with no significant effect on the incidence of other cancers.

It has been suggested that these supplements may have pro-oxidant effects whereas antioxidants consumed in foods may be more balanced. It is also likely that fruits and vegetables contain other substances such as phytochemicals (beneficial plant chemicals) which have cancer preventive action.

8.8 Folate and cancer

Folate is required for *de novo* synthesis of purines and deoxythymidylate which are components of DNA molecules. Deficiency in folate results in uracil misincorporation into DNA and futile DNA repairs leading to chromosomal breakage. The intake of at least five servings of fruit and vegetables daily has been consistently associated with a decreased incidence of cancer. Fruits and vegetables are an excellent source of folate. Folate supplementation has been shown to decrease proliferation of colon mucosal cells in patients with recurrent adenomatous polyps of the colon. However, a recent meta-analysis of folic acid intervention trials did not show harm or benefit in cancer incidence.

8.9 Polyunsaturated Fatty Acids (PUFAs): Beneficial or harmful?

An association between excess saturated fat intake and cancer incidence is supported by epidemiological studies, which have suggested that diets consisting of a low intake of saturated fatty acids and ω-6 PUFAs reduce cancer risk. This is supported by evidence from the traditional Mediterranean diet which is rich in fruits and vegetables and oleic acid derived mainly from olive oil (a plant food). In addition, olive oils contain phytonutrients such as polyphenols which may also interfere with the risk of cancer.

Key points on dietary PUFAs and cancer:

- Arachidonic acid, which is mainly produced from dietary linoleic acid, is the most common ω-6 PUFA in Western diet and cooking fats
- In Western diet, animal foods are the main sources of oleic acid
- Epidemiological studies suggest a dietary pattern to reduce cancer risk is low intake of saturated fatty acids and ω-6 PUFAs
- In animal studies, ω-6 PUFAs have a strong mammary tumour-enhancing effect
- The dietary source of oleic acid is mainly olive oil (a plant food) in the Mediterranean region
- Olive oils are a rich source of omega-9 fatty acids and certain phytonutrients such as polyphenols which may interfere with the risk of cancer
- Dietary linoleic acid in Western animal foods and cooking fats are metabolised to arachidonic acid, which is the most common ω-6 PUFA

- A typical Western diet has 10–20 fold higher amount of ω-6 fatty acids as compared to ω-3 fatty acids, and high intake of proteins
- Intake of plant-based foods provides a balance of ω-3:ω-6 ranging from 1:1 to 1:4
- Health benefits of ω-3-fatty acids have long been thought to be due to their anti-inflammatory properties
- Suppressing inflammation is implicated to be beneficial in reducing cancer risk

In contrast, dietary linoleic acid in Western animal foods and cooking fats are metabolised to arachidonic acid, which is the most common ω-6 PUFA. A typical Western diet has 10–20 fold higher amount of ω-6 fatty acids as compared to ω-3 fatty acids and high intake of proteins. Increased intake of plant-based foods can provide a balance of ω-3:ω-6 ranging from 1:1 to 1:4, which has been reported to be beneficial.

Key points on components of a traditional Mediterranean diet:

- A traditional Mediterranean diet consisting of large quantities of fresh fruits and vegetables, nuts, fish and olive oil, low intake of meat and sugar coupled with physical activity has been associated with reduced risk of heart disease, certain cancers, diabetes, Parkinson's and Alzheimer's diseases
- This diet also contains protective substances such as selenium, glutathione, a balanced ratio of n-6/n-3 FA, high amounts of fibre, antioxidants (e.g. resveratrol from wine and polyphenols from olive oil), vitamins C and E, some of which have been shown to be associated with lower risk of cancer and have cardioprotective effects

Most data regarding the effects of dietary ω-6 PUFAs on human cancers are observational and do not demonstrate cause-effect relationship. In animal studies, ω-6 PUFAs have a strong mammary tumour-enhancing effect. This is possibly due to the effects of prostaglandins and leukotrienes produced from arachidonic acid, which are lipid mediators of inflammation. Inflammation has been linked to the development and growth of cancers. Since chronic inflammation is linked to cancer, diets which can suppress inflammation is implicated to be beneficial in reducing cancer risk. The proposed mechanisms of action of EPA and DHA are as follows: (1) EPA and DHA interactions with lipid rafts generate less potent eicosanoid metabolites, which do not promote tumour development; (2) EPA and DHA can generate resolvers of inflammation such as resolvins, preventing the transition to a diseased state; (3) EPA and DHA modulate immune response e.g. by inducing the secretion of IL-10 but not the pro-inflammatory cytokines, TNF and IL-6, from monocytes.

Key points on immune effects of fish oil, EPA and DHA:

- *In vitro* studies show that fish oil supplements or consumption of fish provides antineoplastic benefits, such as apoptosis of cancer cells, anti-proliferation, and anti-angiogenesis
- EPA and DHA interactions with lipid rafts generate less potent eicosanoid metabolites, which will not promote tumour development
- EPA and DHA can generate resolvers of inflammation such as resolvins, preventing the transition to a diseased state
- EPA and DHA modulate immune response e.g. by inducing the secretion of IL-10 but not the pro-inflammatory cytokines, TNF and IL-6, from monocytes

---

The health benefits of ω-3 fatty acids have long been thought to be due to their anti-inflammatory properties. *In vitro* studies show that fish oil supplements provide anti-neoplastic benefits, such as apoptosis of cancer cells, anti-proliferation, and anti-angiogenesis. However, the benefits of ω-3 fatty acid supplementation for reducing cancer risk or treatment are still uncertain. Epidemiological studies on the impact of fish and fish oil consumption on risk of colon cancer are inconsistent. Recent human data on fish and fish oil consumption suggested that ≥3 years of fish oil supplementation reduced colorectal cancer risk by 49%. Another study demonstrated that fish oil supplementation may be useful in cancer patients who are undergoing chemotherapy and have reduced neutropenia. However, some studies have found an association between increased blood levels of ω-3 fatty acids and increased risk of prostate cancer. Thus, the benefits of fish oil supplementation for cancer risk reduction are still unclear.

Key points on epidemiological studies on PUFAs and cancer:

- Epidemiological studies on the impact of fish and fish oil consumption on risk of colon cancer are inconsistent
- Recent human data on fish and fish oil consumption suggested that ≥ 3 years of fish oil supplementation reduced colorectal cancer risk by 49%
- Fish oil supplementation may be useful in cancer patients who are undergoing chemotherapy and have reduced neutropenia
- However, epidemiological studies have found an association between increased blood level of ω-3 fatty acids and increased risk of prostate cancer

8.10 Dietary fibre and cancer risk reduction

Dietary fibres are thought to reduce cancer risk. High fat and low fibre intake are implicated in cancers such as colon cancer. The high fat intake increases the level of bile acid in the gut leading to the generation of bile acid metabolites which are carcinogenic. High fibre diet can increase stool bulk and speed up transit time of stools, which decrease exposure of mucosa to possible carcinogens and thus reduce absorption of carcinogens. Certain fibres can bind carcinogens. If no dietary fiber is present in the colon, anaerobic bacteria draw their energy from protein fermentation. This metabolism leads to the production of toxic and potentially carcinogenic compounds (such as ammonia or phenolic compounds). High fibre diet has also been shown to protect against colon cancer in a chemical-induced colon cancer rat model. In human studies, high dietary fibre consumption has been associated with lower risk for colorectal adenomas. Dietary fibre intake was reported to reduce risk of colorectal cancer and recurrent adenoma in prostate, lung, colorectal, and ovarian cancers in a screening trial with 55,000 human participants.

Key points on colon cancer and dietary fibre:

- High fat and low fibre intake implicated
- High fat intake increases the level of bile acid in the gut; bile acid metabolites are carcinogens
- High fibre diet increases stool bulk and speeds up transit time resulting in decreased exposure of mucosa to possible carcinogens
- Certain fibres can bind carcinogens
- High fibre diet has also been shown to protect against colon cancer in a chemical-induced colon cancer rat model
- In human studies, high dietary fibre consumption has been associated with lower risk for colorectal adenomas
- Dietary fibre intake reduced risk of colorectal cancer and recurrent adenoma in prostate, lung, colorectal, and ovarian cancer screening trial with 55,000 human participants

---

8.11 Effects of dietary fibre on immunity and gut microbiota

Animal studies have shown that consumption of high fibre content also increased the proportion of $CD8^+$ T and $CD4^+$ T cells and increased NK cell activity in the lamina propria and peripheral blood. Dietary fibre may also alter gut microbiota. High fibre consumption was shown to promote/increase growth of gut *Bifidobacterium. Bifidobacterium annalis* subsp.

Lactis have been found to exert anti-inflammatory activity with production of nitric oxide. Dietary fibres serve as substrates for short-chain fatty acids (SCFAs). Among SCFAs, butyrate and propionate are reported to exhibit strong anti-inflammatory properties by inhibiting TNF-α, IL-8, IL-10, and IL-12 in immune and colonic cells.

Key points on effects of dietary fibre and gut microbiota:

- Increased proportion of CD8$^+$ T and CD4$^+$ T cells and increased NK cell activity in the lamina propria and peripheral blood
- Promote/increase growth of gut *Bifidobacterium*, with an increase in the concentrations of SCFAs
- SCFAs such as butyrate and propionate have strong anti-inflammatory properties by inhibiting TNF-α, IL-8, IL-10, and IL-12 in immune and colonic cells

---

8.12 Effects of dietary fats and proteins on gut microbiota

Diets rich in fats have been associated with cancer risk because fat increases synthesis of bile acids in the liver. These bile acids travel from the enterohepatic circulation into the colon. Once in the colon, the primary bile acids are converted into secondary bile acids, namely, deoxycholic acid and lithocholic acid by certain gut bacteria. Studies have associated increased faecal secondary bile acids from patients with colon polyps and colon cancer.

Diets rich in meat and low in fibre promote proteolytic fermentation by gut *Clostridium*, *Bacteriodes* and *Proteobacteria* resulting in production of nitrogen metabolites such as ammonia and branched SCFAs such as isobutyrte, isovaleric acid and 2-methylbutyric acid which are inflammatory and may enhance colon cancer risk.

8.13 Natural products/compounds and cancer

A number of potential natural compounds as anticancer agents have been documented. These include carotenoids, vitamin D, vitamin E, curcumin from turmeric, reservatrol from red grape skin, limonene from orange or lemon peels, allicin from fresh garlic, astragalus from Chinese herbs, parthenolides from the herb Feverfew, genistein from Red Clover, and proanthocyanidins which are present in many foods such as grape seed, apples, cranberries and tea. Foods such as flaxseed, cordyceps, Maitake or Shiitake mushrooms, and bee propolis may also have potential benefits. Numerous reports on *in vitro* and animal cancer models have been reported for these natural compounds. For example, *in vitro* studies with breast and gastric cancer cells have shown that the sulfur-based amino acids in garlic such as S-allylmercapto-L-cysteine have been found to increase glutathione levels, detoxify carcinogen, inhibit production of ROS, suppress DNA adduct formation, regulate cell cycle arrest and induce apoptosis of cancer cells. It should be noted that some of the natural compounds can interact with conventional chemotherapeutic drugs and advice from the oncologist should be considered if these natural compounds are consumed.

8.14 How do micronutrients enhance tumour immunity?

The key effector mechanisms for the killing of cancer cells involve NK cells and cytotoxic T lymphocytes. Information on the characteristic features and functions of these cells are elaborated in Chapter 1.

Many studies have focused on the role of micronutrients as antioxidants and limited studies are established on their role as enhancers of tumour immunity. The role of vitamins and minerals are also important. The proposed mechanism of cancer risk reduction by antioxidant vitamins are described in Section 8.7. Deficiencies in vitamins and minerals compromise the immune status and this will diminish immune responses against cancerous cells. The mechanisms by which deficiencies in vitamins and minerals diminish immune function are elaborated in Chapters 2 and 3.

Vitamin A is essential for growth and differentiation of cells, vision, immunity and reproduction. Culturing of immature dendritic cells with retinoic acid (vitamin A metabolite) increased migratory ability of dendritic cells *in vitro* and when injected into tumours in animals. This increase in expression of matrix metalloproteinase-9 which degrades the extracellular matrix by retinoic acid can increase the migration of tumour-infiltrating dendritic cells to the draining lymph node, which can boost tumour-specific T cells.

Vitamin D is important for the calcification of bones and maintaining immunity, neural function and pancreatic function. Epidemiologic studies have associated vitamin D deficiency with increased risk of cancers such as colon, prostate, and breast cancer. However, increased risk of pancreatic cancer was observed with higher levels of dietary vitamin D intake.

There are conflicting reports on the role of vitamin E in cancer. Vitamin E supplementation has been shown to increase lymphocyte proliferation and more B cells in non-smokers as compared to placebo, suggesting that vitamin E may have immune protective effects. A two-week supplementation of vitamin E to patients with advanced colorectal cancer showed an increase in CD4:CD8 ratio and enhanced production of Th1 cytokines, IL-2 and IFN-γ, thus suggesting the enhancement of cell-mediated immunity.

Selenium plays an important role in the antioxidant defense network similar to vitamin E by protecting cell membranes from oxidation as it acts as a cofactor for antioxidant selenoproteins. Cancer risk is mitigated with selenium supplementation in selenium-deficient individuals at Linxian province in central north China where soils in this province are selenium-deficient and high incidences of upper gastrointestinal cancer have been reported. Meta-analysis of nine randomised controlled clinical trials showed that supplementation with selenium reduced risk of cancers by 24%. A European Prospective Investigation into Cancer and Nutrition study involving 520,000 participants across 10 Western European countries reported that higher levels of selenium were associated with reduced risk of colorectal cancer. In healthy humans, a 6-month intake of selenium increased T cells by 65% and NK cell cytotoxicity by 56%. Selenium-deficient animals were found to have reduced number of lymphocytes in thymus, NK cell activity and phagocytic activity of neutrophils and macrophages towards infectious agents. A 6-week dietary supplementation of selenium in mice resulted in activation of IL-6 and IFN-γ pathways. Thus, selenium can have protective functions against tumours not only due to its antioxidant effect, but also due to effects on the innate and adaptive immune system.

However, results from the selenium and vitamin E prevention trial (SELECT) which enrolled over 35,000 men showed that there was increased risk of developing prostate cancer. Neither selenium nor vitamin E influenced the development of recurrent colorectal adenoma in 2286 participants who had one or more recurrent colorectal adenomas.

Zinc is largely available in animal proteins. Zinc is involved in the cell cycle, apoptosis and immune cell function. Zinc deficiency affects development of immune cells. Conflicting reports have been published on the benefits of zinc supplementation. Intake of dietary zinc has been associated with decreased risk of colon cancer but has also been negatively associated with gastric, oesophageal and colorectal cancer.

Plasma concentrations of zinc should not exceed 30 μM. Supplementation of zinc should be based on levels present in patients. A cautionary note should be taken as high concentrations of zinc can inhibit immune cell function.

8.15 Immunonutritional support for cancer patients

One of the pathophysiological changes in cancer is metabolic stress which can be assessed by inflammatory markers. Pro-inflammatory cytokines produced by the tumour disrupt the metabolism of carbohydrates, fats and proteins and also affect the neuroendocrine control of appetite resulting in fatigue, muscle wasting and impaired physical activity. Thus, the inflammatory response in cancer inhibits nutrient utilisation and promotion of catabolism. This will result in alteration in body composition and decline in physical function. The prevalence of malnutrition in cancer patients ranges from 20–70%. Nutritional strategies to deal with catabolism have been considered and these include oral nutritional supplementation with essential amino acids or high dose leucine, fish oil, arginine and nucleotides. Physical activities are also included in the European Society for Parenteral and Enteral Nutrition (ESPEN) 2016 guidelines on nutrition in cancer patients. For cancer patients undergoing surgery, early oral nutritional supplementation is the preferred mode of nutrition. For those who are nearing end of life, palliative care is recommended. Food and hydration will also need to be provided as primary care to support comfort and quality of life.

Summary of Chapter 8

Nutrients with antioxidant properties are thought to be important for reducing cancer risk. A diet rich in fresh fruits and vegetables and low intake of meat is highly favourable in reducing cancer risk. Vitamin C has been long known to prevent cancer due to its antioxidant property. Vitamin E, β-carotene and selenium also have antioxidant effects. Based on current evidence available, high doses of a single micronutrient are not recommended. It is still unclear if ω-3 fatty acid supplementation is beneficial for cancer risk reduction or treatment as contradicting results have been reported. A number of natural compounds such as curcumin, reservatrol, limonene, allicin, astragalus, parthenolides, genistein and proanthocyanidins and foods such as flaxseed, cordyceps, Maitake or Shiitake mushrooms and bee propolis may also have potential anticancer effects. Finally, the effects of nutrients in enhancing tumour immunity are summarised. Nutritional strategies for cancer patients include oral nutritional supplementation with essential amino acids or high dose leucine, fish oil, arginine and nucleotides. Physical activity is included in the ESPEN 2016 guidelines on nutrition in cancer patients.

Chapter 9

Exercise immunology

Learning objectives

After studying this chapter, you should be able to:

1. Describe the immune changes in exercise
2. Explain how exercise has anti-inflammatory effects
3. Highlight the implications of exercise for prevention and treatment of chronic inflammation-associated diseases
4. Explain the benefits of immunonutritional support for elite athletes

9.1 Health impact of moderate exercise

Regular moderate exercise is generally useful for health and well-being. Examples of moderate intensity exercise include ballroom dancing, brisk walking and doubles tennis. In general, light-moderate exercise has significant health benefits as it can improve cardiorespiratory fitness, musculoskeletal fitness, flexibility and body composition. The large muscles move repetitively during light-moderate exercise and this increases the use of oxygen and promotes cardiovascular health. In general, regular exercise is beneficial and can reduce risk of chronic diseases such as obesity, stroke, heart disease, high blood pressure and type II diabetes. Exercise has anti-inflammatory effects and therefore regular exercise and good dietary management reduce risk of chronic diseases such as obesity, stroke, hypertension, type II diabetes and cardiovascular diseases.

9.2 Health impact of intense, prolonged or exhaustive physical exercise

Vigorous exercise includes running, singles tennis, aerobics, jumping rope and jogging. If vigorous exercise is intense, prolonged or exhaustive, it can have a negative impact as it results in physiological changes such as increased pro-inflammatory cytokine production, and high rates of protein catabolism occurring with muscle damage, chronic oxidative stress and immune suppression. It has been reported that exercise-induced immunosuppression is mild and transient. Overtraining syndrome in athletes can be harmful as a higher risk of upper respiratory tract infection (URTI) is promoted by intensive, prolonged and exhaustive exercise. Training is also interfered by acute respiratory infections such as sore throats and flu-like symptoms and may lead to poor performance. There has been an immense interest in studying the effects of exercise on the immune system. It should be noted that the effects depend on the intensity, duration and nature of the exercise as well as the fitness level and age of the individual. Highest degree of immune impairment post-exercise occurs in individuals who undertake continuous, prolonged (> 1.5 hour), moderate to high intensity exercise (50–77% maximum oxygen uptake) without food intake. Susceptibility to infection varies in individuals due to factors such as immunocompetence, stress tolerance, recovery strategies and non-training stress factors.

Key points on exercise and hormones:

During exercise, the primary hormones released are

- adrenaline
- noradrenaline
- cortisol
- growth hormone and glucagon

These hormones are also secreted when the body is reacting to emotional and mental stress.

Key points on cortisol:

Cortisol is a hormone produced by adrenal gland. Excess cortisol can lead to

- suppression of immunity
- reduced levels of growth hormone
- decrease in insulin sensitivity
- increased insulin resistance
- reduction in kidney function
- hypertension
- reduced connective tissue strength

A decrease in strength and performance of athletes can be due to chronically elevated levels of cortisol.

The anti-inflammatory effects of intense exercise in elite athletes with high training and competition loads can contribute to partial immunosuppression and increased susceptibility to common infections. Other factors that contribute to immunosuppression are psychological and environmental stress, lack of sleep and inadequate diet. Another possibility is that these elite athletes have an increased exposure to pathogens due to the following reasons:

1. Increased depth and rate of breathing during exercise result in enhanced exposure to airborne bacteria and viruses
2. Saliva secretion falls during exercise and can reduce secretion of anti-microbial substances such as lysozyme, IgA, and α-amylase
3. During prolonged exercise in the heat, gut permeability is increased, thus allowing penetration of bacterial endotoxins into the blood
4. Skin abrasions in contact sports increase risk of infections
5. Close proximity to large crowds
6. Increased frequency of air travel for sporting events

9.3 Exercise and hormones

It is well established that the primary hormones released during exercise are:

- Adrenaline (raises heart rate and blood flow to the muscles)
- Noradrenaline (increases blood pressure)

- Cortisol (suppression of immunity, reduced growth hormone levels, decreased insulin sensitivity, elevated insulin resistance, reduction in kidney function, hypertension, and reduced connective tissue strength)
- Growth hormone and glucagon (maintain normal blood glucose levels and mobilise alternative energy sources for fuel)

These hormones are also secreted when the body is reacting to mental and emotional stress. When cortisol is chronically elevated, strength and performance of athletes are reduced.

9.4 Benefits of training exercise for performance

Exercise training is an adaptive process. Maximum effectiveness of training involves adaptation of muscle to stress. The purpose of physical training is to improve performance. This can only happen when the body can adapt to the stress of physical exercise and improve the capacity to exercise. Overtraining and muscle injury occur when the stress cannot be tolerated.

Elevated oxidative stress occurs resulting in increased oxidant defence via upregulation of antioxidant enzymes such as superoxide dismutase and glutathione peroxidase. Free radicals such as reactive oxygen species and nitric oxide play an important signalling role for metabolism, energy production and capillarisation of muscle fibres. The metabolic effect on the skeletal muscles is increased glycolysis and lipid oxygenation, which is accompanied by a higher storage potential for energy substrates, in particular, glycogen. The net effect is beneficial as there is a general shift of carbohydrate and lipid metabolism in skeletal muscles towards higher storage of glycogen, which is an energy substrate.

9.5 Inflammation, stress and muscle damage

Muscle damage can result as a consequence of acute bouts of intense prolonged exercise and changes in immune response. Reactive oxygen and nitrogen species from oxidative stress and pro-inflammatory cytokines produced in intensive exercise contribute to muscle damage. The muscle is a source of IL-6 production during intensive exercise. Release of molecules such as histamine and prostaglandins causes

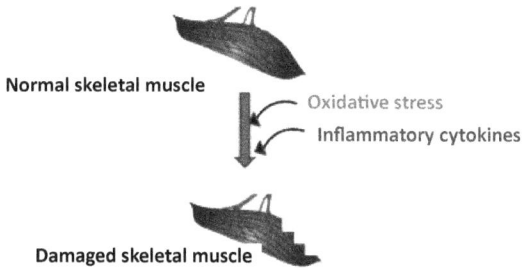

Figure 9.1   Skeletal muscle damage in exercise. Excessive inflammation and oxidative stress contribute to damaged skeletal muscles.

oedema and pain. Activated macrophages and pro-inflammatory cytokines induce muscle cell apoptosis and necrosis, leading ultimately to muscle damage (Figure 9.1).

9.6   Immune changes in exercise

9.6.1   *Circulating cytokine changes*

The immune changes described are either from animal studies, human blood measurements or *in vitro* effects from immune cells isolated from the blood of subjects undergoing exercise.

Key points on immune effects of intense and prolonged exercise:

- Increase in circulating IL-1β and TNF-α (1–2 fold) and IL-6 (over 100-fold)
- Decrease in cytokine inhibitors (IL-1 receptor antagonist and soluble TNF receptors) and the anti-inflammatory cytokine IL-10; severe exercise may result in enhanced susceptibility to infections
- Decrease in circulating numbers of Th1 cells, reduced T cell proliferative responses and reduction in antibody production

---

An acute bout of heavy exercise influences the level of cytokines in the blood (Figure 9.2). During acute heavy exercise, the first cytokine released into the circulation is IL-6 and this is likely due to stimulation by cortisol. The level of IL-6 then declines whilst anti-inflammatory cytokine IL-10 and IL-1 receptor antagonist increase exponentially and persist for a longer duration. IL-1 receptor antagonist blocks IL-1 receptor and thereby blocks the effects of IL-1. Pro-inflammatory cytokines, TNF-α

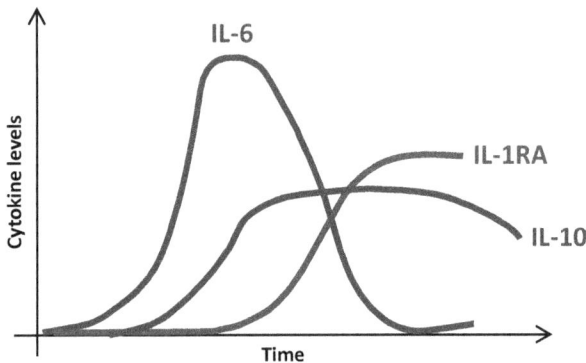

Figure 9.2　Increase in circulating IL-6 in acute bouts of heavy exercise. During acute heavy exercise, the increase in circulating IL-6 occurs rapidly before the increase in anti-inflammatory cytokines, IL-10 and IL-1 receptor antagonist.

and IL-1β, in general do not increase in acute exercise. The number of circulating leukocytes (mainly lymphocytes and neutrophils) increases in acute exercise. However, the magnitude of the increase depends on the intensity and duration of exercise.

Intense and prolonged exercise such as after a marathon results in increase in IL-6, IL-1β and TNF-α in blood, and decrease in cytokine inhibitors (soluble TNF-receptor and IL-1 receptor antagonist) and the anti-inflammatory cytokine IL-10. Concentrations of plasma IL-1β and TNF-α increase by 1–2 fold whilst IL-6 can increase by over 100-fold. Reducing the levels of IL-6 and cortisol would be expected to reduce exercise-induced suppression of immune function in prolonged strenuous exercise. It should be noted that light intensity exercise is not associated with a dramatic increase in circulating IL-6.

Another effect of acute and chronic resistance exercise (exercising the muscles by using an opposing force i.e. dumb bells or resistance band) is a decrease in expression of Toll-like receptors (TLRs) on monocytes. Reduced expression of TLRs can result in subsequent inhibition of pro-inflammatory cytokine production. Prolonged exercise also results in a decreased induction of cytokines and co-stimulatory molecules. Furthermore, both expression of TLR4 on monocytes and inflammatory cytokine production decrease following chronic exercise. The decrease in TLRs and inflammatory capacity of leukocytes could

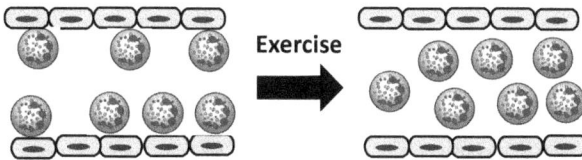

Figure 9.3 Neutrophil demargination. Neutrophils demarginate by loosely attaching to the endothelial cell lining, which results in increased numbers circulating in blood (neutrophilia).

dampen chronic inflammation and potentially reduce the risk of developing chronic diseases. The reason for the decrease in TLR expression is still unclear.

9.6.2 *Effects of exercise on neutrophils*

Acute exercise results in a rapid and dramatic increase in neutrophil numbers (neutrophilia) (Figure 9.3). A few hours later, there is a second, delayed increase in numbers of blood neutrophils. The magnitude of the increase is related to both the duration and intensity of exercise. Hormonal changes also occur in exercise. The initial rapid increase in neutrophils is likely due to adrenaline-induced demargination of the neutrophils. The later increase may be due to release of neutrophils from the bone marrow, which is induced by cortisol. Neutrophils from individuals who undergo an acute bout of exercise have diminished killing capacity (reduced oxidative burst and diminished degranulation responses).

9.6.3 *Effects of exercise on monocytes*

Acute exercise results in a transient (~2 hours) increase in blood monocytes. There is a preferential mobilisation of non-classical macrophages which are $CD14^{low}/CD16^{+}$-expressing monocytes. Animal studies have shown enhanced phagocytosis, increased release of reactive oxygen species and increased chemotaxis.

Key points on effects of moderate exercise:

- Upregulation of pro-inflammatory cytokines and downregulation of anti-inflammatory cytokines IL-10, IL-6 and IL-1 receptor antagonist

- Increase in blood monocytes
- Increase in NK cell cytotoxicity
- Biphasic transient increase in blood lymphocytes (lymphocytosis)
- Decrease in the percentage of circulating Th1 cells but has little effect on Th2 cells
- No change in serum immunoglobulin levels
- Increase in salivary IgA production and secretion rate

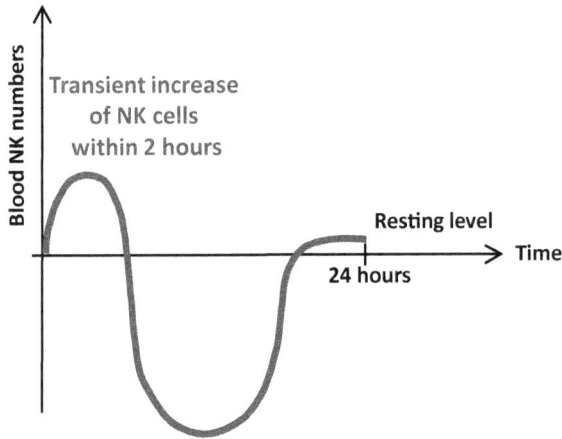

Figure 9.4   Effect of exercise on NK cell numbers in blood. Acute exercise induces transient increase in circulating NK cell numbers which decrease and then return to normal within 24 hours.

9.6.4 *Effects of exercise on natural killer (NK) cells*

Acute exercise results in transient increase in NK cell numbers in the blood, which decrease to less than half of normal levels for approximately 2 hours and then return to normal resting values within 24 hours (Figure 9.4).

Animal studies revealed that regular exercise can enhance cytotoxic action of NK cells. However, whether this also occurs in humans in response to regular exercise is still unknown.

9.6.5 *Effects of exercise on B cells*

Short or prolonged exercise either does not change or slightly increases serum immunoglobulin concentration. Short duration of moderate exercise

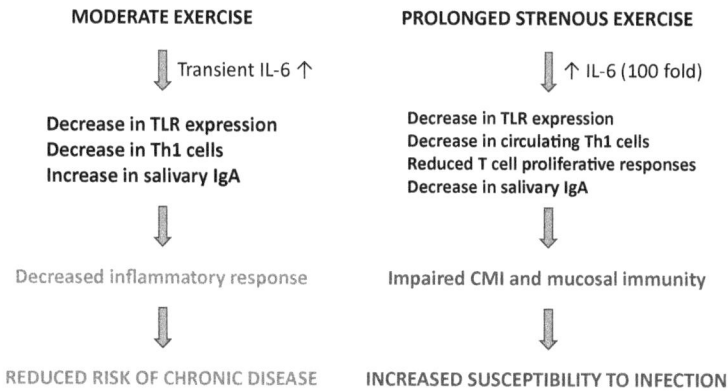

MODERATE EXERCISE	PROLONGED STRENOUS EXERCISE
⇓ Transient IL-6 ↑	⇓ ↑ IL-6 (100 fold)
Decrease in TLR expression **Decrease in Th1 cells** **Increase in salivary IgA**	Decrease in TLR expression Decrease in circulating Th1 cells Reduced T cell proliferative responses Decrease in salivary IgA
⇓	⇓
Decreased inflammatory response	Impaired CMI and mucosal immunity
⇓	⇓
REDUCED RISK OF CHRONIC DISEASE	INCREASED SUSCEPTIBILITY TO INFECTION

Figure 9.5   Comparison of immune alterations in moderate versus prolonged strenuous exercise.

results in a moderate increase in salivary IgA. In contrast, prolonged high intensity exercise results in a decrease in salivary IgA (Figure 9.5). It is likely that the increases in salivary IgA observed after moderate exercise training may contribute to the reduced susceptibility to URTIs associated with regular moderate exercise. Similarly, the increased susceptibility to URTIs in high-performance endurance athletes undergoing intensive training is consistent with the lower concentrations of salivary IgA and secretion.

9.6.6 *Effects of exercise on T cells*

Acute exercise causes transient biphasic changes in the numbers of circulating lymphocytes. During and immediately after exercise, there is a dramatic increase in T lymphocytes in blood. During the early stages of recovery, the number of lymphocytes decreases to pre-exercise levels and then they steadily restore to resting values. These changes are related to changes in cytokines released and increase in stress hormones, particularly cortisol. Adrenaline released in exercise induces increase in β_2-adrenergic receptors expression on T cells. The alterations in cytokine and adhesion molecule expression can also contribute to the alterations in haemopoiesis and increased mobilisation of cells from the bone marrow.

In conclusion, prolonged intensive exercise has been found to cause a decrease in TLR expression on monocytes, dramatic rise in circulating IL-6 levels, decrease in the percentage of circulating Th1 cells with little change in numbers of Th2 cells, and decrease in salivary IgA (Figure 9.5) leading to impaired cell-mediated immunity (CMI) and mucosal immunity. It is well established that Th1 cells promote cell-mediated immune responses. The decrease in Th1 cells is likely due to increased levels of stress hormones during exercise and mobilisation of less mature leukocytes from the bone marrow to the blood circulation. There is a reduction in T cell function and a decrease in stimulated antibody synthesis in elite athletes undergoing a period of intensive training. This could be a contributing factor to increased susceptibility to URTIs. In addition, the reported symptoms could be caused by airway inflammation due to the inhalation of dry air, pollution, or drying out of mucosal surfaces during intense prolonged exercise. Finally, the increase in susceptibility to infection can also be due to a variety of other stressors such as physical, environmental and psychological stress.

In contrast, moderate exercise results in decreased inflammatory response and is associated with reduced risk of chronic inflammatory diseases. IL-6 is now considered as both a pro-inflammatory and anti-inflammatory cytokine. It stimulates the synthesis of anti-inflammatory cytokines such as IL-1 receptor antagonist and IL-10. IL-6 also induces synthesis of acute phase proteins which are considered as markers of inflammation. Some of the acute phase proteins such as serum amyloid A promote inflammation by recruiting immune cells to the inflammatory sites whilst others such as anti-trypsin have anti-inflammatory properties. Thus, regular moderate exercise is beneficial.

9.7 Impact of exercise in chronic metabolic diseases

Exercise is an inducer of lipolysis. Thus, exercise can induce the breakdown of fats. A healthy diet and exercise result in a decrease in classical macrophages (M1 macrophages) in adipose tissue and reduced accumulation of lipids and macrophages in liver and skeletal muscles, ultimately resulting in reduced inflammation (Figure 9.6). M1-type macrophages produce TNF, IL-6 and nitric oxide, whereas M2-type macrophages produce anti-inflammatory cytokines. Thus, M1 macrophages are considered as inducers of inflammation whilst M2

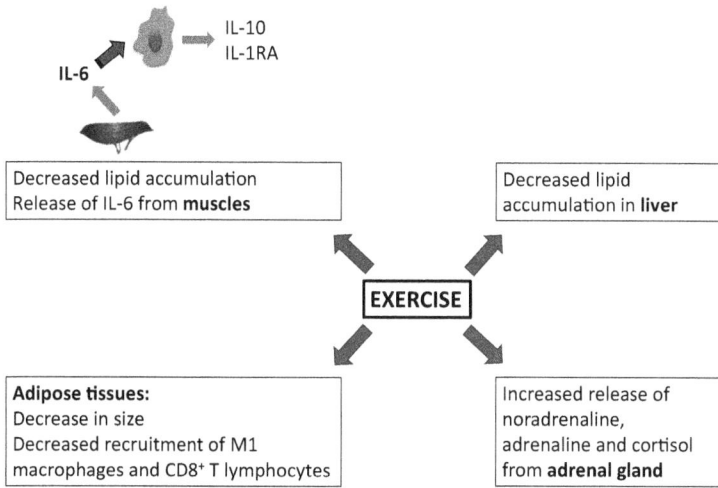

Figure 9.6   Effects of exercise on adrenal glands, muscles, liver and adipose tissue.
Exercise results in release of adrenaline, noradrenaline and cortisol from adrenal glands, decreased lipid accumulation in liver and muscles, production of IL-6 from muscles and anti-inflammatory cytokines from monocytes, decrease in adipose tissue size and decrease in recruitment of M1 macrophages and CD8$^+$ T lymphocytes into adipose tissue.

macrophages reduce inflammation. Regular exercise increases energy expenditure and burns off fat, which can cause reduction in visceral fat and induction of anti-inflammatory effects. Exercise also improves cardiovascular health and thus can lower risk of developing chronic heart disease. Observational epidemiological studies have also shown an association between high levels of physical activity and a reduced risk of developing type II diabetes.

9.8   Exercise and cancer

Inflammation and obesity have been linked to the pathogenesis of cancer. Exercise appears to have anti-inflammatory effects and therefore regular moderate exercise is implicated to protect against chronic diseases such as cancer. Comprehensive reviews by the World Cancer Research Fund and the International Agency for Research on Cancer have reported protective effects of exercise on postmenopausal breast cancer and colon risk and also in endometrial, lung and pancreatic cancers. In addition, animal model data show a positive effect of exercise on increased clearance of lung metastases by macrophages. These studies also show that exercise training resulted in enhanced *in vivo* mechanisms of natural immunity, increased *in vitro* NK cell cytotoxicity, and reduced pulmonary tumour metastases. Thus, regular exercise is beneficial in reducing cancer risk and enhancing tumour immunity.

9.9 Immunonutritional support for elite athletes

9.9.1 *Benefits of immunonutritional support for elite athletes*

Elite athletes are individuals involved in competitive sports usually at a national or international level. The major benefits of immunonutritional support for elite athletes are to provide adequate energy and boost immunity (Table 9.1). To reduce infection risk and boost immune function, a number of nutritional strategies can be implemented.

9.9.2 *High carbohydrate intake*

An adequate intake of carbohydrate and protein is required to ensure adequate energy. High carbohydrate intake has been shown to limit metabolic stress by maintaining blood glucose levels and lowering circulating stress hormone and anti-inflammatory cytokines.

Carbohydrate beverages can also help to prevent dehydration and maintain saliva flow during exercise. Saliva secretion falls during exercise and can reduce secretion of antimicrobial substances such as lysozyme, IgA, and α-amylase.

Table 9.1   Summary of immunonutritional support for elite athletes.

The most widely used immunonutritional supports are vitamin and mineral supplements and proteins, such as whey for supplements or isolated amino acids.

The benefits of immunonutritional support for elite athletes are to:

1) Boost immunity
2) Provide adequate energy, carbohydrate and protein intake and avoid deficiencies of micronutrients to maintain health

---

(*Continued*)

Table 9.1   (*Continued*)

The outcomes of nutritional supplements for elite athletes include:

1) Improvement of performance
2) Strengthening of immune function
3) Minimisation of exercise recovery period

Key points on high carbohydrate intake:

• Limits metabolic stress by maintaining blood glucose
• Lowers circulating stress hormone and anti-inflammatory cytokines
• Carbohydrate beverages help prevent dehydration and maintain saliva flow during exercise

---

9.9.3 *Supplementation with antioxidants, vitamin C, vitamin E and trace minerals*

Elite athletes commonly undergo intensive training for a few days or weeks at certain times of the season. Following relatively short periods (1–3 weeks) of intensified training, there are significant reductions in neutrophil function, lymphocyte proliferation, secretory IgA, and IFN-β-producing T cells. Thus, with prolonged strenuous training, both innate and adaptive immunity are decreased but they are not clinically immune-deficient. However, they are at risk of infections. Therefore, elite athletes can benefit from immunonutritional support to improve performance, boost immunity and minimise exercise recovery period. The most commonly used supplements are vitamins and minerals to maintain a healthy immune system. One proposed mechanism for antioxidant supplements is their action on reducing IL-6 release from the muscle fibers of the exercising legs. The reduction in IL-6 and hence reduction in cortisol response would be expected to limit the mild, transient decrease in immune function induced by exercise. This could explain the reported lower incidence of URTIs in ultramarathon athletes supplemented with vitamin C (alone or in combination with other antioxidants).

There is some evidence that regular intake of high dose vitamin C or E can reduce stress response to prolonged exercise. However, there is risk of adaptation to training. In addition, a delay in healing and recovery of strength and increase in oxidative stress after muscle-damaging exercise have been reported in regular high dose vitamins C and E supplementation. Thus, excessive supplementation is not recommended in healthy individuals. A balanced diet with plenty of fruits and vegetables is the best option as the body develops an endogenous antioxidant defense which improves with exercise.

There are certain situations, such as during high-altitude training periods, whereby vitamin C and vitamin E supplementation are probably advantageous. Hypoxia is a condition where there is lack of oxygen and this is experienced by people living at high altitudes. Hypoxia can induce intensive free radical production and weaken the body's defence system.

There is insufficient evidence that deficiencies in minerals such as zinc, magnesium and iron in elite athletes undergoing high intensity exercise aggravate exercise-induced immune responses.

Key points on antioxidant vitamin supplementation:

- A balanced diet with plenty of vegetables and fruits is the best option
- Body develops an endogenous antioxidant defense which improves with exercise
- Some evidence that regular intake of high dose vitamin C or E can reduce stress response to prolonged exercise
- However, there is risk of adaption to training
- Exercise-induced vasodilation and angiogenesis for capillarisation of muscle fibres are prevented by acute supplementation of vitamin C and vitamin E
- Excessive supplementation is not recommended in healthy individuals
- A delay in muscle healing and strength recovery, and increased oxidative stress after muscle-damaging exercise have been reported in elite athletes supplemented with regular high doses of vitamin C and vitamin E
- However, supplementation is probably advantageous in certain situations; these include high-altitude training periods, where hypoxia intensifies free radical production and weakens the body's defence system

---

The use of proteins such as whey for supplements or isolated amino acids has been reported in many studies. An adequate intake of carbohydrates and proteins is required to ensure adequate energy. Evidence is also emerging that some nutritional supplements including *Lactobacillus* probiotics and flavonoids such as quercetin can augment immune function and reduce the incidence of infections in exercise-stressed athletes. However, there is limited data to support the use of megadoses of vitamin C, α-glucans, γ-3 polyunsaturated fatty acids, echinacea, ginseng, or bovine colostrums for elite athletes.

9.9.4 *L-glutamine supplementation*

L-glutamine has been used as a nutritional supplement for elite athletes. It can increase the concentration of L-glutamine in the liver and muscles. This can then increase the concentration of glutathione (GSH), leading to improved antioxidant capacity and thereby reduce oxidative stress. Glutamine is one of the components of GSH which is a tripeptide consisting of L-glutamine, L-cysteine and glycine that functions as the major antioxidant and redox buffer against oxidative damage. Besides its antioxidant effect, L-glutamine can reduce muscle damage and inflammation by promoting uptake of water and sodium ions, and release of potassium ions, which increase hydration and volume and thereby resist muscle damage induced by exhaustive exercise. L-glutamine is also required by rapidly proliferating cells providing nitrogen for purine and pyrimidine nucleotide synthesis for new DNA and RNA production. L-glutamine is also an essential amino acid and a substrate for gluconeogenesis. Despite a good rationale for glutamine supplementation, laboratory-based animal studies have been disappointing. It has been suggested that there is sufficient glutamine available in the body stores. However, there is evidence that glutamine supplementation can reduce the incidence of exercise-induced URTIs.

Key points on glutamine supplementation:

- Used as a nutritional supplement for elite athletes
- Increases concentration of L-glutamine in the liver and muscles which in turn increases concentration of GSH, which reduces oxidative stress (GSH functions as the major antioxidant)
- Increases hydration and volume
- Can reduce muscle damage and inflammation induced by exhaustive exercise
- Required for rapidly proliferating cells
- Can decrease the incidence of exercise-induced URTIs

9.9.5 *Whey supplementation*

Milk contains two fractions: whey and caseins. Whey contains bovine serum albumin, immunoglobulins (e.g. IgA), peptides, amino acids, vitamins and minerals, lactoperoxidase and β-lactoglobulin, α-lactalbumin, lactoferrin and lactoferricin. Lactoperoxidase, immunoglobulins and peptides synergise to exert anti-viral and anti-bacterial activity. Lactoferrin and lactoferricin also have anti-microbial activity. Whey has an impact on the immune system as it contains components such as L-glutamine, which can act as immunomodulator and is able to attenuate oxidative stress in immune cells.

Key points on whey supplementation:

- Milk contains two fractions: whey and caseins
- Components of whey include immunoglobulins (e.g. IgA), β-lactoglobulin, lactoferrin, lactoperoxidase, peptides, amino acids, vitamins and minerals
- Lactoperoxidase, immunoglobulins and peptides synergise to exert anti-viral and anti-bacterial activity
- Lactoferrin and lactoferricin also have anti-microbial activity
- Amino acids include L-glutamine, L-arginine, L-lysine, cysteine and taurine
- L-glutamine acts as an immunomodulator and has the ability to attenuate oxidative stress

---

9.9.6 *Other supplements*

Other nutritional supplements including *Lactobacillus* probiotics and flavonoids such as quercetin have been reported to augment immune function and diminish the incidence of infections in exercise-stressed athletes. However, there is limited data to support the use of minerals, vitamin E or megadoses of vitamin C, ω-3 polyunsaturated fatty acids, β-glucans, bovine colostrums, and nutriceuticals such as ginseng and echinacea for elite athletes.

Summary of Chapter 9

Regular physical activity for the general population can reduce risk of chronic diseases such as type II diabetes, heart disease, stroke, high blood pressure, obesity and osteoarthritis. The benefits of regular exercise on health are apparent even when the exercise is of light-moderate intensity such as brisk walking for an hour each day. However, intense, prolonged or exhaustive exercise can have negative health impacts due to changes such as increased pro-inflammatory cytokine release, high rates of protein catabolism, oxidative stress, muscle damage and mild, transient immune suppression. It should be noted that the effects depend on the intensity, duration and nature of the exercise as well as the fitness level and age of the individual. Immune impairment is highest when continuous prolonged moderate to high intensity exercise is performed without food intake. Increased susceptibility to URTIs has been reported in athletes who undergo exhaustive exercise. IL-6 is the first cytokine released into the circulation during acute bouts of exercise which then declines, whilst anti-inflammatory cytokine IL-10 and IL-1 receptor antagonist increase exponentially and persist for a longer duration. Acute heavy exercise induces a biphasic transient increase in blood lymphocytes (lymphocytosis), decrease in T cell function and production, no change in serum immunoglobulin levels but decreases in salivary IgA concentration and secretion rate. Light intensity exercise is not associated with dramatic increase in circulating IL-6. Moderate exercise is beneficial in chronic metabolic syndrome as it can induce breakdown of fat, reduce accumulation of lipids and macrophages in liver and skeletal muscles, as well as decrease M1 macrophage activity in adipose tissue, ultimately resulting in reduced inflammation. In animal models of cancer, exercise can enhance macrophage function, NK cell cytotoxic function, and clearance of tumours. Antioxidant vitamins, glutamine and whey supplementation have also been found to be beneficial as immunonutritional support for elite athletes.

Chapter 10

Immunonutrition in the elderly and critically ill patients

Learning objectives

At the end of studying this chapter, you should be able to:

1. Explain the theory of aging, immunosenescence, inflammaging and immune modelling
2. Explain the nutritional strategies to mitigate oxidative stress
3. Describe the changes in the immune system of the elderly
4. State the known micronutrient deficiencies in the elderly and effects of supplementation
5. Explain the immune response in critical illness
6. Describe the roles of glutamine, arginine, nucleotides, PUFAs and probiotics/prebiotics on immunonutrition in critical illness
7. Explain the role of immunonutrition in burn patients

10.1 Introduction

10.1.1 *What is aging?*

Aging is a process of becoming older accompanied by physiological changes that lead to senescence or a progressive decline in the biological functions of different tissues, organs or systems including the immune system.

The age-related decline in immunity is referred to as immunosenesence. A decline in immune function leads to an increased susceptibility to infections and frailty, diminished response to vaccination and increased susceptibility to age-related inflammatory diseases such as autoimmune diseases, cardiovascular diseases, metabolic syndrome, Type II diabetes, Alzheimer's disease and cancer. Low grade chronic inflammation is thought to contribute to age-related inflammatory diseases and this has been referred to as 'inflammaging'. The terms 'immunosenescence' and 'inflammaging' have given a negative connotation to aging. A new paradigm referred to as 'immune remodeling' has recently emerged. The new paradigm suggests that aging leads to modified responses of the immune system making it more adapted to cope with challenges such as pathogens and autoantigens.

10.2 Oxidative stress in aging

There are many theories on aging. The free radical theory has been established since 1956. It speculates that the body's defence system is not able to deal with reactive oxygen species (ROS)-induced damage. The cumulative oxidative damage leads to cell death and affects life span. More recently, the theory of 'oxidative stress hypothesis" has been proposed and is widely accepted. It states that *'oxidative damage is not solely triggered by the unrestricted ROS production, but is also caused by other oxidants, such as reactive lipid species and reactive nitrogen species (RNS)'*. Oxidative stress is caused by the imbalance of antioxidant and pro-oxidant levels (redox status). The generation and oxidation of ROS in a normal healthy individual occurs in a controlled manner. However, when there is an imbalance of redox status and the immune system is dysregulated, this may lead to systemic inflammation. The unresolved chronic inflammation leads to degenerative diseases such as cardiovascular, cancer, diabetes, dementia, osteoporosis, arthritis and metabolic syndrome. There are antioxidant enzymes that protect against oxidative stress in the body. These include catalase, glutathione peroxidase, and superoxide dismutase. Non-enzymatic ROS scavengers are available in our diet such as β-carotene, vitamin C and vitamin E. These mechanisms can help to protect against oxidative stress in the body.

10.3 Interventions to mitigate oxidative stress

Dietary intake of antioxidants can reduce the risk of age-related diseases. The vitamins and trace minerals that have antioxidant functions are as follows:

- Vitamin C, also known as ascorbic acid, is a cofactor for many enzyme-catalysed reactions to maintain the integrity of connective and vascular tissues and to maintain the cell structure by enhancing collagen biosynthesis. The process of collagen synthesis involves the hydroxylation of lysine and proline with vitamin C acting as a cofactor. This process is required for collagen synthesis in order to maintain the structure of the cells. Severe vitamin C deficiency causes scurvy which is a serious disease. The defect in collagen synthesis leads to bleeding gums, loss of teeth, discoloration of the skin and wounds that do not heal. Vitamin C is also required for iron absorption and enhances lymphocyte proliferation. In addition, vitamin C protects against oxidative damage due to ROS generated by phagocytes in response to bacterial and viral infections.
- γ-tocopherol has potent antioxidant activity and is the primary form of vitamin E.
- Carotenoids (lutein and zeaxanthin) form the pigment in the macula. Vitamin A supplementation can reduce oxidative stress and age-related macular degeneration. In addition, T cell proliferation and cytotoxicity are enhanced.
- Trace minerals with antioxidant properties include magnesium, copper and selenium.

10.4 Age-related changes in the immune system

Immunosenescence affects both innate and adaptive immunity. There is a decrease in the function of neutrophils, macrophages, natural killer (NK) cells, dendritic cells, T and B cells. The number of neutrophils remains unchanged but chemotaxis and phagocytosis are reduced. It is unclear whether the production of ROS is increased or decreased due to conflicting results. As for the monocytes, the numbers remain unchanged but macrophage precursors in the bone marrow decline. Antigen presentation is reduced due to reduced expression of MHC Class II molecules.

Phagocytosis and production of superoxide anions and cytokines are reduced. There is an increase in NK cells but there is a difference in NK subsets. A decline in CD56^{bright} NK cells and increase in CD56^{dim} NK cells has been observed. This leads to a decrease in NK cell cytotoxicity. Other changes include a decrease in cytokine and chemokine production, proliferation and expression of cytokine receptors.

In the elderly, there is a reduction in plasma of naïve T and B cells while the number of memory T cells increases. This is thought to be due to thymic involution and decline in bone marrow function, resulting in reduced number of naïve lymphocytes. There is also loss of function and diversity of T and B cell repertoire, which contribute to reduced ability to respond to new antigens and vaccines. This is thought to be the reason for the increased susceptibility to infections. Lifelong and chronic exposure to pathogens result in several rounds of T cell replication and they become late-differentiated effector memory T cells with low proliferative activity. Increase in memory T cells was observed in older people infected with the human cytomegalovirus but these memory CD8$^+$ T cells do not express costimulatory molecule CD28 which is required for T cell activation. Late stage memory senesecent T cells have suppressive activity and are producers of pro-inflammatory cytokines. In the elderly as compared to young adults, blood levels of pro-inflammatory cytokines such as tumour necrosis factor (TNF) and IL-6 are higher, thus, contributing to low grade chronic inflammation which is also referred to as 'inflammaging'. Variations in environmental factors such as stress, exercise, diet and nutrition as well as genetics can contribute to the onset and progression of immunosenescence in the elderly. Dietary or nutritional intervention to mitigate inflammation or to boost immunity appears to be a rational and amenable approach. Due to heterogeneity in the elderly population, there is difficulty in making appropriate dietary recommendations as a whole. The health problems associated with aging vary between individuals. Thus, nutrition plans need to be individualised to address specific nutrition deficiencies.

Key points on dysfunction of immunity in aging:

- Reduced neutrophil chemotaxis and phagocytosis
- Monocyte numbers remain unchanged but their precursors in the bone marrow decline

- Reduced antigen presentation
- Reduced NK cell cytotoxicity
- Reduced number of naïve T and B cells
- Increase in memory T and B cells, and late-differentiated senescent T cells
- Reduced ability to respond to new antigens
- Reduced diversity of T and B cell repertoire

10.5 Micronutrient deficiencies in the elderly

A total of 37 articles were included in the pooled systematic analysis which was published in 2015. The report revealed that of the 20 nutrients analysed, deficiency in 6 vitamins and trace minerals were considered to be a possible public health concern: vitamin D, riboflavin, thiamine, calcium, magnesium and selenium. The effects of deficiency in some of these vitamins and trace elements are shown in Table 10.1.

Table 10.1   Micronutrient deficiency and impact of supplementation in the elderly on immune function.

Vitamins	Effects of Deficiency	Effects of Supplementation in Deficient Elderly
Vitamin A	Impaired vision Impaired immunity Increased risk of neurodegeneration	Reduced oxidative stress and age-related macular degeneration; augmented T cell cytotoxicity and proliferation
Riboflavin (Vitamin B_2)	Increased oxidative stress	Lowered plasma homocysteine levels
Vitamin E	Low plasma levels found in patients with dementia and Alzheimer's disease Reduction in cellular immunity leading to increased susceptibility to viral infections	Improved NK cytotoxic activity, neutrophil chemotaxis, phagocytosis, enhanced mitogen-induced lymphocyte proliferation and IL2 production

Table 10.1 (*Continued*)

Vitamins	Effects of Deficiency	Effects of Supplementation in Deficient Elderly
Vitamin C	Predisposition to infections	Increased neutrophil chemotaxis and phagocytosis Increased antibody levels Increased absorption of dietary iron
Vitamin B$_{12}$ and folate	Can lead to anaemia, cognitive decline and neurological problems Low Vitamin B$_{12}$ and high homocysteine can impair immunity and increase inflammation	Correction of anaemia
Vitamin D	Increased risk of falls and fracture	Enhanced immunity by stimulating phagocytosis by macrophages and protecting immune cells against apoptosis
Iron	Can lead to anaemia, reduced phagocytic activity and NK numbers, lower lymphocyte response, reduced delayed-type hypersensitivity response, and lower production of IFN-β	Overcomes iron deficiency anaemia, stimulates lymphocyte proliferation and differentiation
Calcium	Increased risk of falls and fracture	Can decrease dietary iron absorption
Zinc	Impaired immunity: decrease in mitogen-induced proliferation of lymphocytes, decrease in production of Th1 cytokines	Improved cell-mediated immunity, reduced susceptibility to infections
Copper	Impaired immunity	Reduced susceptibility to infections
Magnesium	Susceptibility to stress, defective membrane functions, inflammation, cardiovascular diseases, diabetes and immune dysfunction	Has anti-inflammatory effect as indicated by reduced C-reactive protein levels
Selenium	Suppression of immune function	Increase in CD4$^+$ T cells and NK cells

10.6   Impact of micronutrients, probiotics and prebiotics on immunity in the elderly

10.6.1   *Vitamin E and vitamin C*

Vitamin E is a lipid-soluble antioxidant present in the membrane of all nucleated cells and is particularly abundant in membranes of immune cells. Animal studies have shown that vitamin E can enhance T cell function. Many studies have reported that vitamin E supplementation is beneficial for healthy elderly subjects. Increase in production of IL-2 and activation-induced T cell proliferation was observed.

Vitamin C is a water-soluble antioxidant which can counteract age-related oxidative stress. It has also been shown to be involved in enhanced antibody generation and differentiation, and maturation of T cells and NK cells.

10.6.2   *Zinc*

Zinc supplementation improved cell mediated immunity. Increased delayed-type hypersensitivity response and increased expression of IL-2 mRNA were reported. Varied results were reported on the lymphocyte phenotypes and response to vaccination. Randomised, double-blind placebo-controlled trials showed that zinc supplementation for 12 months resulted in lower incidence of infections and lower levels of plasma TNF and oxidative stress markers in the elderly. In another double-blind randomized controlled trial, supplementation of zinc in the elderly resulted in increased levels of activated T helper and cytotoxic T cells.

10.6.3   *Selenium*

Studies in aged people show that supplementation of selenium increases $CD4^+$ T cells that persisted for 2 months following supplementation discontinuation. In other studies of aged people, the selenium concentration was positively associated with an enhancement in the number of NK cells. Selenium is present in selenoproteins. They are antioxidant enzymes such as glutathione peroxidase, thioredoxin reductase, and iodothyronine deiodinases which are important in controlling oxidative stress. Although it appears rational to administer selenium to mitigate oxidative stress, there are controversies as selenium has been reported to increase the risk of neuronal diseases due to neurotoxic effects.

10.6.4 *Probiotics*

Probiotics has been defined as *'live microorganisms that provide beneficial effects by modulating the immune function in the gastrointestinal tract and distant tissues via the mucosal immune system'*. Immune cells that traffic to and from the mucosal sites are critical in modulating immunity. The most characterised probiotics belong to members of the genera *Bifidobacterium, Lactobacillus* and *Streptococcus*. Elderly individuals with reduced beneficial gut microbes were found to have decreased antigen-specific secretory IgA in intestine. Thus, it is rational that elderly people would benefit from probiotic intake. Studies on probiotic consumption in the elderly have shown increased phagocytosis and bactericidal activity as well as increased number of NK cells and their tumour-icidal activity.

10.6.5 *Prebiotics*

Prebiotics consist of indigestible dietary fibres which are food for 'beneficial' gut bacteria. Supplementation with a prebiotic galactooligosaccharide has been shown to increase NK cell activity with reduction in pro-inflammatory cytokines such as IL-1, TNF and IL-6, as well as increased IL-10 by peripheral blood mononuclear cells and increased phagocytosis by neutrophils. There was also an increase in beneficial gut bacteria such as *Bifidobacteria*. However, administration of healthy older adults with prebiotics did not improve response to influenza and pneumo-coccal vaccines.

10.6.6 *Impact of Mediterranean diet on aging*

A traditional Mediterranean diet is one that is rich in olive oil, vegetables, fruits, unrefined cereals, legumes, low in meat and dairy products, and with a moderate alcohol intake. It has been widely accepted that this diet and lifestyle is associated with decreased incidence of chronic diseases such as cardiovascular diseases. This diet has also been associated with increased longevity such as those residing in the Mediterranean region.

10.7 Examples of health problems associated with micronutrient deficiencies

Lower levels of folate, vitamins A, B_{12}, C and E have been reported in the plasma of patients with dementia and Alzheimer's disease. Supplementation with these vitamins may be potentially beneficial for these patients. Studies have also revealed a positive relationship between improved hearing threshold and intake of antioxidants and magnesium. Elevated serum homocysteine levels have been implicated as a risk factor for osteoporosis, and supplementation with vitamin B_{12} and folic acid reduced plasma homocysteine levels. Increased homocysteine levels have been associated with decline in physical function in elderly adults. Thus, there is a potential benefit of folic acid supplementation in these elderly people. Folic acid supplementation is also potentially beneficial for prevention of stroke in patients with cardiovascular disease by lowering homocysteine levels.

10.8 Immunonutrition for critically ill patients

10.8.1 *Introduction*

Critically ill patients are defined as '*acutely ill patients with altered organ function such that homeostasis cannot be maintained without medical intervention in intensive care units, such as mechanical ventilation, vasoactive support for haemodynamics and renal replacement therapy*'. These patients include those with cardiovascular disease, burns, acute respiratory distress, severe trauma/injury and/or sepsis. The systemic inflammatory response which occurs as a result of stresses such as in trauma/injury, infection or surgery may exert high metabolic demand upon the body and lead to depletion of essential nutrient stores. The initial immune response is the release of pro-inflammatory cytokines that orchestrate the host response to injury and infection. However, these cytokines induce high levels of inflammation which can lead to immunosuppression. It is common that immune dysfunction occurs in these patients. It may be possible to modulate the immune response for reducing patient morbidity and mortality.

In patients undergoing major surgery, there is an initial pro-inflammatory acute phase response followed by immunosuppression. Monocyte and macrophage activities are suppressed. There is also a reduction in total number of T cells, cytotoxic T cells, and NK cells.

In trauma cases, there is a dramatic alteration in the immune response. The function of monocytes and secretion of IL-1β which play an important role in protection against infections are decreased.

In sepsis, complex pathophysiological alterations occur. The inflammatory response including ROS and RNS release and activation of the coagulation and complement cascade systems is activated as summarised in Figure 10.1. Mediators of inflammation such as eicosanoids, platelet activating factor, nitric oxide, vasoactive amines, kinins and cytokines are also involved. Elevated plasma levels of chemokines (CCL2/MCP-1, MIP-1β), IL-8, IL-6, and IL-10 have been reported. The systemic inflammatory response involves the endothelium, platelet aggregation, and smooth vascular and bronchial muscles. This response may lead to impairment of the microcirculation, vascular permeability, coagulation, pulmonary gas exchange and substrate utilisation, which may affect organ function.

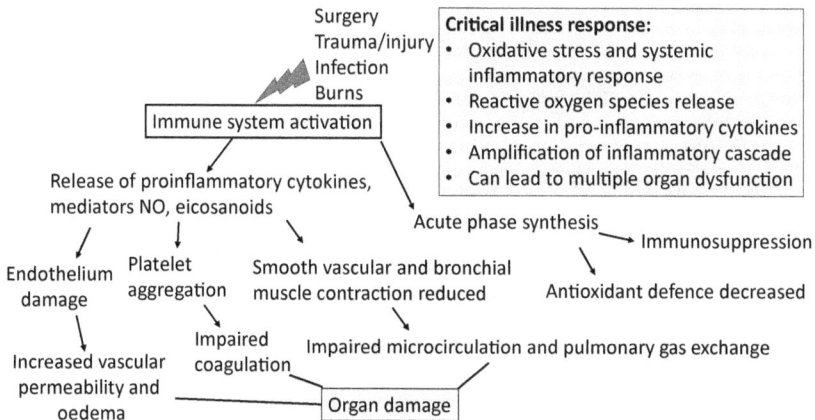

Figure 10.1   Overview of immune response to critical illness.

Thus, there is potential for alleviating the severity of the inflammatory response by supplying nutritional substrates as highlighted in Figure 10.2. Nutritional substrates can also modulate cell-mediated immunity and humoral response in the critically ill and restore their immune system to defend against invading pathogens. Early enteral feeding (intake of food through the mouth or through a tube that goes directly to the stomach or small intestine) of critically ill patients can also maintain the structure and function of the intestinal mucosa to serve as mucosal barrier against invading pathogens.

In critically ill patients, there is potential scope for nutritional therapy as the acute illness-associated malnutrition can lead to muscle wasting and delayed wound healing. These patients are also at higher risk of infections and longer hospital stays.

In severe critically ill patients requiring mechanical ventilation, such as those with septic shock, acute respiratory distress syndrome, and major burns and trauma, oxidative stress occurs. Oxidative stress is defined as an imbalance between increased ROS and RNS and endogenous antioxidant mechanisms. The circulating levels of primary antioxidant defense micronutrients, in particular copper, selenium, zinc, and vitamins C and E are decreased. The 2009 European Society for Parenteral and Enteral Nutrition (ESPEN) guidelines state that the full range of vitamins and minerals are

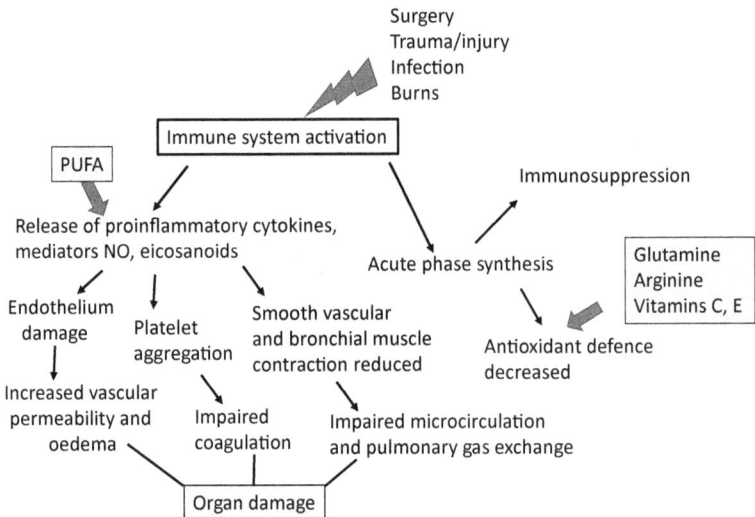

Figure 10.2   Immunonutrition to modulate immune response to critical illness.

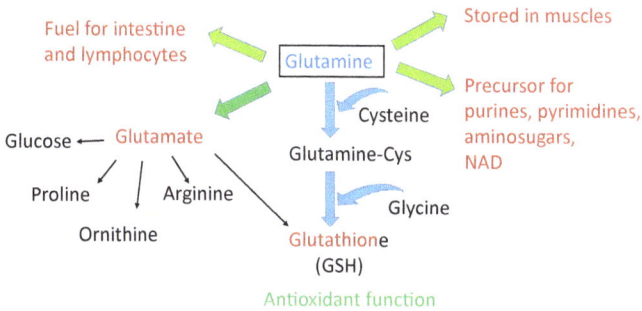

Figure 10.3   Overview on the role of glutamine.

Key points on the role of glutamine:

- Most prevalent free amino acid in body
- > 60% are in muscles
- Major metabolic fuel for cells in gastrointestinal tract and rapidly proliferating cells (immune cells)
- Nucleotide precursor
- Antioxidant: GSH detoxifies ROS
- Attenuates iNOS expression
- Restores mucosal IgA
- Augments NK activity

---

integral to providing nutritional support. The uses of glutamine, arginine, n-3 PUFAs and nucleotides for critically ill patients are more complicated.

10.8.2 *Glutamine supplementation*

Glutamine is the most abundant amino acid in the blood and muscle cells. It is used for the biosynthesis of proteins and source of nitrogen for the synthesis of purines, pyrimidines, nucleotides, amino sugars and glutathione (GSH), an antioxidant, as shown in Figure 10.3.

10.8.3 *Effects of glutamine on immunity*

In the gut, GSH is involved in the detoxification of ROS. Experimentally induced intestinal GSH deficiency results in impaired mucosal integrity and cell function. Glutamine represents the major metabolic fuel for cells of the gastrointestinal tract (enterocytes, colonocytes) and rapidly proliferating cells such as T lymphocytes. Glutamine supplementation has been reported to enhance upper respiratory tract immunity by restoring mucosal IgA as well as augment the activity of NK cells.

A marked depletion of muscle intracellular glutamine has been reported in patients after major bouts of trauma and injury, burns, infections, surgery and pancreatitis. Enteral glutamine therapy has been reported to be effective in preventing infections in non-infected patients with severe multiple trauma and reduction in incidence of pneumonia, bacteraemia and severe sepsis. However, for intensive care patients who are already severely stressed or infected, enteral supplementation is inadequate and it is likely that parallel parenteral supplementation is required.

Enteral glutamine as part of an immunonutrition formula has been recommended for patients with traumatic brain injury or perioperative patients. However, due to the lack of clear benefits of glutamine for patients in intensive care, Society of Critical Care Medicine (SCCM)/ American Society of Parenteral and Enteral Nutrition (ASPEN) guidelines do not recommend parenteral glutamine supplementation. This applies particularly for patients diagnosed with early-stage multisystem organ failure or ongoing shock requiring vasopressor support.

10.8.4 *Arginine supplementation*

Arginine is a dibasic amino acid and is involved in many metabolic pathways within the human body. It is a precursor for proteins, polyamines, glutamate, histidine and creatine. Arginine via the formation of glutamate yields proline and hydroxyproline which is required for synthesis of connective tissues. Arginine is an important component of the urea cycle, which is the only pathway for the elimination of toxic ammonia from the body. Ornithine, which is a by-product of the urea cycle, is a precursor for polyamines which are essential for cell proliferation and differentiation. Arginine is required for the synthesis of creatine which is required for muscle contraction. Arginine stimulates the release of hormones such as growth hormone, insulin growth factors, insulin and glucagon as shown in Figure 10.4. Arginine is the only substrate for the generation of nitric oxide (NO) which is important for vasodilation, neurotransmission and immunity.

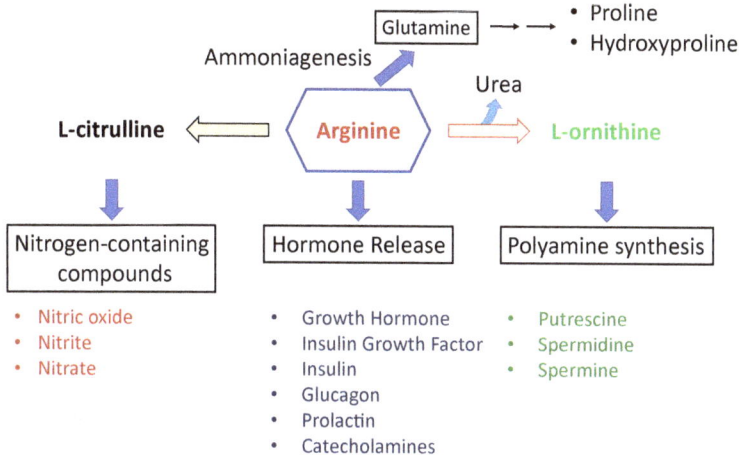

Figure 10.4 Overview of the effects of arginine.

Key points on the effects of arginine:

- Amino acid derived from diet or via urea cycle
- Metabolised within enterocyte via arginase to form urea and ornithine
- Yields increased amounts of proline and hydroxyproline via glutamate
- Precursor of polyamines for cell proliferation
- Stimulates release of hormones
- Only substrate for production of NO which is important for vasodilation, neurotransmission and immunity
- Reverses increased intestinal mucosalpermeability due to inhibition of NO synthesis in experimental models of ischaemia-reperfusion intestinal injury
- Supplementation in trauma patients improved T cell function and phagocytosis
- May enhance systemic inflammatory response due to enhanced NO release leading to cytotoxicity of cells, hypotension due to vascular dilation, and impaired coagulation

---

10.8.5 *Effects of arginine on immunity*

Increased lymphocyte and monocyte proliferation were observed in enterally arginine-fed healthy individuals and surgical and intensive care unit patients. Experimental animal and human studies on supplementation with arginine showed increased phagocytosis rate and improved T cell function in trauma. However, clinical trials with critical ill patients have failed to demonstrate improvements in patient outcomes. In patients with sepsis or severe inflammatory response syndrome, it was unclear whether arginine may enhance systemic inflammatory response due to increased NO release. This could lead to impaired coagulation and vasodilatation leading to refractory hypotension. Thus, arginine supplementation should be undertaken with caution if administered to critical patients such as those with systemic inflammatory response syndrome or organ failure.

10.8.6 *Nucleotides*

Nucleotides are key components of RNA and DNA. In healthy individuals, nucleotides are absorbed from the diet. The absence of nucleotides in the diet results in loss of helper T lymphocytes and suppression of IL-2 production. The demand for nucleotides increases during episodes of infection following trauma and injury to allow for increased need of immune cells to synthesise more nucleotides. Experimental studies on removal of dietary nucleotides showed impaired mucosal integrity and function. This implies that nucleotides are critical in intestinal function. The critical function of nucleotides in immunity is supported by impaired T cell function, reduced NK activity, suppressed lymphocyte proliferation, reduced IL-2 production, reduced phagocytosis and elimination of pathogens when nucleotides were removed from the diet.

10.8.7 *n-3 Polyunsaturated fatty acids*

The effects of n-3 PUFAs are explained in Chapter 4. Though fish oil supplementation is potentially beneficial in reducing inflammatory response and exerting anti-thrombotic effects, the effect on suppressing T cell function may not be desirable. It has been reported that the immunosuppressive effect contributed to increased lipid peroxidation (may result in cell damage) and decreased antioxidant levels, particularly vitamin E levels. Thus, supplementation of n-3 PUFAs in critically ill patients needs to be undertaken with caution.

10.8.8 *Probiotics*

A meta-analysis of five small randomised clinical trials with an overall of 281 patients showed that probiotics were associated with reduced nosocomial infections and ventilator-associated pneumonia in trauma patients. However, these findings require confirmation as results from meta-analyses of randomised clinical trials with larger samples of critical care patients remain inconclusive.

10.8.9 *Immunonutrition for patients with burns*

The current recommendation for burn care is early enteral nutrition with 1.5–2.5 g/kg/day protein and continuous administration of a standard polymeric formula via a small-bowel feeding tube, and vitamin (vitamins B_1, C, D, and E) and trace mineral (zinc, copper, and selenium) supplementations. The benefits of glutamine for burn care are unclear although it has been reported that there is mortality benefit; more studies are needed.

For patients with severe burns, early enteral nutrition over parenteral nutrition is recommended following the 2016 guidelines from both SCCM/ASPEN and ESPEN. This has been reported to lower the rate of ischemia/reperfusion injury, improved gastrointestinal contractility and reduced intestinal permeability.

Summary of Chapter 10

This chapter is focused on the immunonutritional support for the elderly and critically ill patients. The introduction explains the immune changes in the elderly and gastrointestinal changes that can affect absorption of dietary nutrients. Due to the heterogeneity of the elderly population, there is difficulty in making appropriate recommendations for micronutrient supplementation. Thus, the nutrition plans need to be individualised to address nutrient deficiencies. The main health problems of the elderly include heart disease, loss of vision, hearing loss, osteoporosis, dementia, and Alzheimer's disease. There is evidence that some elderly patients with heart disease have lower blood levels of micronutrients. The relationship between improved hearing threshold and antioxidant and magnesium supplement intake has been documented. There is scientific evidence to show that elderly patients with dementia and Alzheimer's disease have lower plasma levels of folate, vitamins A, B_{12}, C and E. This knowledge is useful to optimise the nutritional status of the elderly to promote their health and to some extent prevent or reduce progression of the disease. Immunonutritional support is also an important subject for critically ill patients. The evidence for glutamine and arginine supplementation to provide a good quality of life is explained in this chapter.

Recommended readings and references

Chapter 1 Elements of the immune system

Abas A, Lichtman A and Pillai S (2015). Cellular and Molecular Immunology, 8th Edition. W. B. Saunders, Philadelphia, USA.

Coico R and Sunshine G (2015). Immunology: A short course. John Wiley & Sons, New York, USA.

Fukatsu K and Kudsk KA (2011). Nutrition and gut immunity. *Surg Clin North Am* 91(4): 755–770.

Izcue A, Coombes JL and Powrie F (2006). Regulatory T cells suppress systemic and mucosal immune activation to control intestinal inflammation, *Immunol Rev* 212: 256–271.

Iwata M, Hirakiyama A, Eshima Y, Kagechika H, Kato C and Song S (2004). Retinoic Acid Imprints Gut-Homing Specificity on T Cells. *Immunity* 21(4): 527–538.

Kelly D and Coutts A (2000). Early nutrition and the development of immune function in the neonate. *Proc Nutr Soc* 59: 177–185.

Keusch GT (2003). The history of nutrition: Malnutrition, infection and immunity. *J Nutr* 133: 336S–340S.

Parham P (2014). The Immune System, 4th edition, Garland Science, New York, USA.

Rescigno M, Lopatin U and Chieppa M (2008). Interactions among dendritic cells, macrophages, and epithelial cells in the gut: implications for immune tolerance. *Curr Opin Immunol* 20(6): 669–675.

Chapter 2 Vitamins and immune function

Bendich A (1993). Nutrition and Immunology, 217–228, in Human Nutrition — A Comprehensive Treatise, edited by Kharlfeld DM, Plenum Press, New York, USA.

Cassani B, Villablanca EJ, De Calisto J, Wang S and Mora JR (2012). Vitamin A and immune regulation: Role of retinoic acid in gut associated dendritic cell education, immune protection and tolerance. *Mol Aspects Med* 33(1): 63–76.

Chua WJ and Hansen TH (2012). Immunology: Vitamins prime immunity. *Nature* 491: 680–681.

Chun RF, Adams JS and Hewison M (2011). Immunomodulation by vitamin D: implications for TB. *Expert Rev Clin Pharmacol* 4(5): 583–591.

Meydani SN, Han SN and Wu D (2005). Vitamin E and immune response in the aged: Molecular mechanisms and clinical implications. *Immunol Rev* 205: 269–284.

Mora JR, Iwata M and von Andrian UH (2008). Vitamin effects on the immune system: vitamins A and D take centre stage. *Nat Rev Immunol* 8(9): 685–698.

Pekmezci D (2010). Vitamin E and immunity. *Vitam Horm* 86: 179–215.

Raverdeau M and Mills KHG (2014). Modulation of T Cell and Innate Immune Responses by Retinoic Acid. *J Immunol* 192: 2953–2958.

Ross CA (2012). Vitamin A and retinoic acid in T cell–related immunity. *Am J Clin Nutr* 96(suppl): 1166S–1172S.

Rytter MJH, Kolte L, Briend A, Friis H and Christensen VB (2014). The immune system in children with malnutrition — a systematic review. *PLoS ONE* 9(8): e105017.

Tengerdy RP (1990). The role of vitamin E in immune response and disease resistance. *Ann NY Acad Sci* 587, 24–33.

Chapter 3 Minerals and immune function

Bonaventura P, Benedetti G, Albarede F and Miossec P (2015). Zinc and its role in immunity and inflammation. *Autoimmun Rev* 14: 277–285.

Haase H and Rink L (2014). Zinc signals and immune function. *Biofactors* 40(1): 27–40.

Hughes DJ, Fedirko V, Jenab M, Schomburg L, Méplan C, Freisling H, Bueno-de-Mesquita HB, Hybsier S, Becker N-P, Czuban M, Tjønneland A, Outzen M, Boutron-Ruault MC, Racine A, Bastide N, Kühn T, Kaaks R, Trichopoulos D, Trichopoulou A, Lagiou P, Panico S, Peeters PH, Weiderpass E, Skeie G, Dagrun E, Chirlaque M-D, Sánchez M-J, Ardanaz E, Ljuslinder I, Wennberg M,

Bradbury KE, Vineis P, Naccarati A, Palli D, Boeing H, Overvad K, Dorronsoro M, Jakszyn P, Cross AJ, Quirós JR, Stepien M, Kong SY, Duarte-Salles D, Riboli E and Hesketh JE (2015). Selenium status is associated with colorectal cancer risk in the European prospective investigation of cancer and nutrition cohort. *Int J Cancer* 136: 1149–1161.

Kieliszek M and Blazejak S (2013). Selenium: Significance, and outlook for supplementation. *Nutr* 29: 713–718.

Kubena KS and McMurray DN (1996). Nutrition and the immune system: A review of nutrient-nutrient interaction. *J Am Diet Assoc* 96: 1156–1164.

Lee DH, Anderson KE, Harnack LJ, Folsom AR and Jacobs Jr DR (2004). Heme iron, zinc, alcohol consumption and colon cancer: Iowa Women's Health Study. *J Natl Cancer Inst* 96: 403–407.

Lee EH, Myung SK, Jeon YJ, Kim Y, Chang YJ, Ju W, Seo HG and Huh BY (2011). Effects of selenium supplements on cancer prevention: Meta-analysis of randomized controlled trials. *Nutr Cancer* 63: 1185–1195.

Li P, Xu J, Shi Y, Ye Y, Chen K, Yang J and Wu Y (2014). Association between zinc intake and risk of digestive tract cancers: A systematic review and meta-analysis. *Clin Nutr* 33, 415–420.

Marikovsky M, Ziv V, Nevo N, Harris-Cerruti C and Mahler O (2003). Cu/Zn superoxide dismutase plays important role in immune response. *J Immunol* 170: 2993–3001.

Percival SS (1998). Copper and immunity. *Am J Clin Nutr* 67: 1064S–1068S.

Qiao L and Feng Y (2013). Intakes of heme iron and zinc and colorectal cancer incidence: a meta-analysis of prospective studies. *Cancer Causes Control* 24, 1175–1183.

Tam M, Gomez S, Gonzalez-Gross M and Marcos A (2003). Possible roles of magnesium in the immune system. *Eur J Clin Nutr* 57: 1193–1197.

Wintergerst ES, Maggini S and Hornig DH (2006). Contribution of selected vitamins and trace elements to immune function. *Ann Nutr Metab* 51: 301–323.

Wood SM, Beckham C, Yosioka A, Darban H and Watson RR (2015). Beta-carotene and selenium supplementation enhances immune response in aged humans. *Integr Med* 2000(2): 85–92.

Chapter 4   Fatty acids and inflammation

Alexander JW (1998). Immunonutrition: the role of omega-3 fatty acids. *Nutrition* 14: 627–633.

Calder P (2012). Omega 3 polyunsaturated fatty acids and inflammation: nutrition or pharmacology? *Br J Clin Pharmacol* 75(3): 645–662.

Grimble RF (1998). Modification of inflammatory aspects of immune function by nutrients. *Nutr Res* 18(7): 1297–1317.

Hokari R, Matsunaga H and Miura S (2013). Effect of dietary fat on intestinal inflammatory diseases. *J Gastroenterol Hepatol* 28(Suppl 4): 33–36.

Nadtochiy SM and Redman EK (2011). Mediterranean diet and cardioprotection: the role of nitrite, polyunsaturated fatty acids, and polyphenols. *Nutrition* 27(7–8): 733–744.

Okada Y, Tsuzuki Y, Ueda T, Hozumi H, Sato S, Hokari R, Kurihara C, Watanabe C, Tomita K, Komoto S, Kawaguchi A, Nagao S and Miura S (2013). Trans fatty acids in diets act as a precipitating factor for gut inflammation? *J Gastroent Hepatol* 28(Suppl 4): 29–30.

Veldhoen M and Brucklacher-Waldert V (2012). Dietary influences on intestinal immunity. *Nat Rev Immunol* 12(10): 696–708.

Chapter 5 Food allergy and food intolerance

Prescott SL, Jennings S, Martino D, D'Vaz N and Johannsen H (2010). Role of Dietary Components in the Epidemic of Allergic Disease, 353–370, in Dietary components and immune function, edited by Watson RR, Zibadi S and Preedy VR, Humana Press, New Jersey, USA.

Chapter 6 Probiotics and prebiotics

Kechagia M, Basoulis D, Konstantopoulou S, Dimitriadi D, Gyftopoulou K, Skarmoutsou N and Fakiri EM (2013). Health benefits of probiotics: a review. *ISRN Nutr* 2013: 481651.

Klaenhammer TR, Kleerebezem M, Kopp MV and Rescigno M (2012). The impact of probiotics and prebiotics on the immune system. *Nat Rev Immunol* 12: 728–733.

Round JL and Mazmanian SK (2009). The gut microbiome shapes intestinal immune responses during health and disease. *Nat Rev Immunol* 9(5): 313–323.

Chapter 7 Autoimmunity and nutrition

Antico A, Tampoia M, Tozzoli R and Bizzaro N (2012). Can supplementation with Vitamin D reduce the risk or modify the course of autoimmune diseases? A systematic review of the literature. *Autoimm Rev* 12: 127–136.

D'Aurizio, Villalta D, Metus P, Doretto P and Tozzoli R (2015). Is Vitamin D a player or not in the pathophysiology of autoimmune thyroid diseases? *Autoimm Rev* 14: 363–369.

Huang S, Wei JC, Wu D and Huang Y (2010). Vitamin B6 supplementation improves proinflammatory responses in patients with rheumatoid arthritis. *Eur J Clin Nutr* 64: 1007–1013.

Kamen DL and Tangpricha V (2010). Vitamin D and molecular actions on the immune system: Modulation of innate and autoimmunity. *J Mol Med* 88: 441–450.

Klack K, Bonfa E and Neto EFB (2012). Diet and nutritional aspects in systemic lupus erythematosus. *Rev Bras Reumatol* 52(3): 384–408.

Chapter 8 Diet nutrition and cancer

Bohnsack BL and Hirschi KK (2004). Nutrient regulation of cell cycle progression. *Ann Rev Nutr* 24: 433–453.

Donaldson MS (2004). Nutrition and cancer: A review of the evidence for an anticancer diet. *Nutr J* 3: 19.

Fonseca-Nunes A, Jakszyn P and Agudo A (2014). Iron and cancer risk — a systematic review and meta-analysis of the epidemiological evidence. *Cancer Epidemiol Biomarkers Prev* 23(1): 12–31.

Gerber M (2012). Omega-3 fatty acids and cancers: a systematic update review of epidemiological studies. *Br J Nutr* 7(Suppl 2): S228–239.

Gratz SW, Richardson AJ, Duncan SH, Russell WR, Fyfe C, Johnstone AM, Flint HJ and Holtrop G (2015). Influence of dietary carbohydrate and protein on colonic fermentation and endogenous formation of N-nitroso compounds. Proceedings of the Nutrition Society, Summer Meeting, 14–17 July 2014, Carbohydrates in health: friends or foes, 74 (OCE1), E4.

Janakiram NB, Mohammed A, Madka V, Kumar G and Rao C (2016). Prevention and treatment of cancers by immune modulating nutrients. *Mol Nutr Food Res* 60: 1275–1294.

Kantor ED, Lampe JW, Peters U, Vaughan TL and White E (2014). Long-Chain Omega-3 Polyunsaturated Fatty Acid Intake and Risk of Colorectal Cancer. *Nutr Cancer* 66: 716–727.

Key TJ (2011). Fruits and vegetables and cancer risk. *Bri J Cancer* 104: 6–11.

Kunzmann AT, Coleman HGC, Huang W-Y, Kitahara CM, Cantwell MM and Berndt SI (2015). Dietary fiber intake and risk of colorectal cancer and incident and recurrent adenoma in the Prostate, Lung, Colorectal, and Ovarian Cancer Screening Trial. *Am J Clin Nutr* 102(4): 881–890.

Meadows GG (2012). Diet, nutrients, phytochemicals, and cancer metastasis suppressor genes. *Cancer Metastasis Rev* 31: 441–454.

Melina V, Craig W and Levin S (2016). Position of the Academy of Nutrition and Dietetics: Vegetarian Diets. *J Acad Nutr Diet* 116(12): 1970–1980.

Moukayed M and Grant WB (2013). Molecular Link between Vitamin D and Cancer Prevention. *Nutrients* 5: 3993–4021.

Rock CL *et al.* (2020). American Cancer Society Guideline for Diet and Physical Activity for Cancer Prevention. *CA Cancer J Clin.* doi:10.3322/caac.21591.

Chapter 9   Exercise immunology

Chazaud B (2016). Inflammation during skeletal muscle regeneration and tissue remodeling: application to exercise-induced muscle damage management. *Immunol Cell Biol* 94: 140–145.

Cruzat VF, Krause M and Newsholme P (2014). Amino acid supplementation and impact on immune function in the context of exercise. *J Internatl Soc Sports Nutr* 11: 6.

Fatiyris UG and Jamaurtas AZ (2016). Insights into the molecular etiology of exercise-induced inflammation: opportunities for optimizing performance. *J Inflam Res* 9: 175–186.

Gleeson M (2013). Nutritional Support to Maintain Proper Immune Status during Intense Training. *Nestlé Nutr Inst Workshop Ser* 75: 85–97.

Gleeson M (2016). Immunological aspects of sport nutrition. *Immunol and Cell Biol* 94: 117–123.

Gleeson M and Pyne DB (2016). Respiratory inflammation and infections in high-performance athletes. *Immunol Cell Biol* 94: 124–131.

Nielsen HG, Oktdalen O, Opsta PK and Lyberg T (2016). Plasma cytokine profiles in long-term strenuous exercise. *J Sports Med* 2016: 7186137.

Ortega E (2016). The "bioregulatory effect of exercise" on the innate/inflammatory responses. *J Physiol Biochem* 72: 361–369.

Pedersen BK and Hoffman-Goetz L (2000). Exercise and the immune system: regulation, integration, and adaptation. *Physiol Rev* 80: 1055–1081.

Walsh NP, Gleeson M, Shephard RJ, Gleeson M, Woods JA, Bishop NC, Fleshner M, Green C, Pedersen BK, Hoffman-Goetz L, Rogers CJ, Northoff H, Abbasi A and Simon P (2011). Position statement. Part one: Immune function and exercise. *Exerc Immunol Rev* 17: 6–63.

Walsh NP and Oliver SJ (2016). Exercise, immune function and respiratory infection: An update on the influence of training and environmental stress. *Immunol Cell Biol* 94: 132–139.

Chapter 10 Aging and critical illness

Alam I, Almajwal AM, Alam W, Alam I, Ullah N, Abulmeaaty M, Razak S, Khan S, Pawelec G and Pracha PI (2019). The immune-nutrition interplay in aging — facts and controversies. *Nutr Health Aging* 5: 73–95.

Albaugh VL, Stewart MK and Barbul A (2017). Cellular and Physiological Effects of Arginine in Seniors, 317–336, in Nutrition and Functional Foods for Healthy Aging, edited by Watson R, Academic Press, Massachusetts, USA.

Cruzat V, Rogero MM, Keane KN, Curi R and Newsholme P (2018). Glutamine: Metabolism and Immune Function, Supplementation and Clinical Translation. *Nutrients* 10: 1564–1595.

Grimble RF (2009). Basics in clinical nutrition: Immunonutrition — Nutrients which influence immunity: Effect and mechanism of action. *eSPEN, Eur e-J Clin Nutr and Metabol* 4: e10–e13.

Maijo M, Clements SJ, Ivory K, Nicoletti C and Carding SR (2014). Nutrition, Diet and Immunosenescence. *Mech Ageing Dev* 136–137: 116–128.

McCarthy MS and Martindale RG (2018). Immunonutrition for Critical Illness: What is the role? *Nutr Clin Pract* 33(3): 348–358.

Montgomery SC, Streit SM, Beebe ML, Pickney J and Maxwell IV (2014). Micronutrient needs of the elderly. *Nutr Clin Pract* 29: 435–444.

Roehl K (2016). Immunonutrition in 2016: benefit, harm or neither? *Pract Gastroenterol XL* 8: 27–39.

Suchner U, Kuhn KS and Furst P (2000). The scientific basis of immunonutrition. *Proc Nutr Soc* 59: 553–563.

Glossary

adaptive immune response	Immune response by antigen-specific T and B lymphocytes.
adenosine triphosphate (ATP)	A high-energy compound made up of purine adenine, ribose and three phosphate units. It is a source of metabolic energy for cells.
adhesion molecules	Cell surface proteins that mediate cells to bind to each other.
alternative complement pathway	One of the pathways of complement that is triggered without involving antibodies.
allergy	An immune response to a substance such as a particular food, pollen, bee venom and household chemicals.
alpha-linolenic acid	An essential fatty acid found in leafy green vegetables, flaxseed oil, soy oil, fish oil and fish products. It is an ω-3 fatty acid.
Alzheimer's disease	A progressive brain disorder that kills brain cells and causes memory loss and cognitive decline.
anaemia	A condition in which there is lack or low numbers of red blood cells or haemoglobin.
antibody	A protein that is produced by plasma cells in response to a foreign substance known as antigen.

(Continued)

antigen	Any molecule or substance that triggers an immune response.
atherosclerosis	A disease in which walls of the artery accumulate deposits of lipids and scar tissue that can build up to impair blood flow.
atopy	A genetic tendency to develop allergic diseases.
autoimmune disease	A condition in which your immune system attacks your body. The pathological damage caused by the adaptive immune response to self-antigen gives rise to autoimmune disease.
B cells	A type of lymphocyte that is responsible for antibody production when it differentiates to a plasma cell. Another name is B lymphocyte.
B7 molecules	Costimulatory molecules present on the surface of antigen-presenting cells such as dendritic cells. They are either B7.1(CD80) or B7.2 (CD86) proteins.
B cell receptor	A membrane-bound immunoglobulin molecule that serves as a receptor for a specific antigen.
C1	Complement protein C1 is the first complement protein activated in the classical pathway of complement activation.
C3	Complement protein C3 is the central component of the complement system.
calcitriol	The primary active form of Vitamin D in the body.
carbohydrate	A macronutrient that is composed of carbon, hydrogen and oxygen derived from plants and provides energy.
cardiovascular disease	An abnormal condition involving dysfunction of the heart and blood vessels that can lead to heart attack or stroke.
catabolism	Breakdown or degradation of larger molecules to smaller molecules.

(Continued)

(Continued)

CCR9	A chemokine receptor that binds CCL25. The CCL25/CCR9 axis is critical for homing of T cells to the gut.
CCR10	A chemokine receptor that binds CCL27 and CCL28 and is expressed on skin-homing T cells.
CD4 T cell	T cell that expresses CD4 on the T cell surface. CD4 acts as a co-receptor in recognition of antigenic peptides presented to the T cell receptor in association with MHC class II molecules.
CD40	A cell surface glycoprotein on B cells that interacts with CD40 ligand on T cells to trigger B cell proliferation, differentiation and antibody isotype switching.
CD40 ligand	A transmembrane protein on T cells that interacts with CD40 on B cells.
CD8 T cell	T cell that expresses CD8 on the T cell surface. CD8 acts as a co-receptor in recognition of antigenic peptides presented to the T cell receptor in association with MHC Class I molecules.
cell-mediated immunity	An adaptive immune response mediated by T and B cells.
cholecalciferol	Vitamin D_3, a form of vitamin D found in animal foods and is the form that is synthesised in the body under sunlight.
classical pathway of complement activation	Activation of complement pathway which is triggered by antibody bound to antigen.
coeliac disease	A disease caused by immune reaction that damages the lining of the small intestine when a person is exposed to allergens such as gluten.
coenzyme	Organic (carbon-containing) component of enzymes. Coenzymes bind to the enzyme and assist in enzymatic activity. Many coenzymes are B vitamins.

(Continued)

cofactor	Small non-protein substance that enhances or is essential for enzyme action. Trace minerals such as iron, zinc, manganese and copper function as cofactors.
commensal	A microorganism that resides in the human body. Normally does not cause disease and can be beneficial.
complement	A group of plasma proteins that act in a cascade of reactions to kill pathogens in the extracellular spaces and in blood.
complex carbohydrate	A nutrient that consist of long chains of glucose molecules such as starch, glycogen and fibre.
co-stimulatory molecule	Cell surface protein on antigen presenting cells that are required for T cell activation. For example, B7.1 on antigen presenting cell binds to CD28 on T cells and serves as Signal 2 for T cell activation. CD40 ligand on T cells serves as co-stimulator when it binds to CD40 on B cells.
Crohn's disease	A chronic inflammatory disease of the small intestine that causes diarrhoea, abdominal pain, rectal bleeding, weight loss and fever.
CTLA4	An inhibitory cell surface protein on T cells that acts as an 'off' switch when it binds to B7 molecules. It functions as an immune checkpoint and downregulates the immune response.
cytokine	A large group of proteins secreted by cells that plays an important role in many cellular processes such as inflammation, cell proliferation, differentiation and chemotaxis. Cytokines bind to specific receptors on their target cells.
dementia	A medical condition with symptoms related to cognitive decline such as difficulties in remembering, thinking, problem solving or language.

(Continued)

(Continued)

dendritic cell	Professional antigen presenting cell that is derived from the bone marrow. It has a branched, dendrite-like morphology and is present in tissues.
dietary fibre	Non-digestible carbohydrate parts of plants.
docosahexaenoic acid (DHA)	A metabolic derivative of alpha-linolenic acid. DHA is a source of D-series of resolvins and protectins which are resolvers of inflammation.
eicosapentaenoic acid (EPA)	A metabolic derivative of alpha-linolenic acid. EPA is the source of E-series of resolvins which are resolvers of inflammation.
ergocalciferol	Vitamin D_2, a form of vitamin D that is found exclusively in plant foods.
essential fatty acids	Fatty acids that must be consumed in the diet because they cannot be made by the body. They are linoleic acid and alpha-linolenic acid.
eosinophil	One of the three types of granulocytes. It contains granules that are secreted when stimulated. Eosinophils are important in the killing of parasitic worms.
eosinophilia	Abnormal increase in number of eosinophils in blood.
epithelium	Structure consisting of cells bound tightly to each other. It can be a single layer of cells such as in the respiratory and gut linings or multiple layers such as in the epidermis.
Fc	Part of the antibody in the carboxy-terminal halves of the two heavy chains of the antibody molecule.
Fc receptor	Cell surface receptor that binds to the Fc part of the antibody molecule.
FcεR	Cell surface receptor on mast cells, basophils and activated eosinophils that bind IgE.
fortified foods	Foods in which nutrients are added that did not originally exist in the food or existed in insignificant amounts.

(*Continued*)

glycogen	The storage form of glucose in humans and animals.
granulocyte	Three types of granulocytes are neutrophil, eosinophil and basophil. They are irregular in shape and have multi-lobed nuclei and cytoplasmic granules.
granzyme	A group of proteolytic enzymes present in the granules of T cells and NK cells.
gut-associated lymphoid tissue (GALT)	Lymphoid tissues associated with the gastrointestinal tract include the Peyer's patches in the intestine and intraepithelial lymphocytes.
haem iron	Iron that is part of haemoglobin and myoglobin, found only in animal foods such as meat, fish and poultry.
heavy chain	The larger of the two polypeptide chains in the antibody molecule.
helper T cell	CD4 T cells that function to help other cells of the immune system.
histamine	An amine that is stored in mast cell granules and is released when mast cells are activated. It causes dilation of local blood vessels and contraction of smooth muscles.
homocysteine	An amino acid that requires adequate levels of Vitamin B_{12}, folate and vitamin B_6 for its metabolism. High levels of homocysteine in blood are associated with increased risk of cardiovascular disease.
humoral immunity	Immunity that is mediated by antibodies.
immunoglobulin	Refers to antibody.
immunosenescence	Age-related progressive decline in the biological function of the immune system.
inflammaging	Low grade chronic inflammation that is related to aging.
inflammation	An immune response to infection, injury and bites that is characterised by redness, swelling and sometimes pain due to the accumulation of fluid, plasma proteins and cells of the immune system at the site of injury or infection.

(*Continued*)

(Continued)

innate immunity	Front line of host defense against invading pathogen with no generation of immunological memory.
integrin	A class of cell-surface glycoproteins that mediates adhesive interactions between cells and the extracellular matrix.
interferon-γ	A cytokine that is produced by Th1 cells, NK cells and CD8$^+$ T cells. The major function is to activate macrophages in innate and adaptive immunity. It differs from IFN-α and IFN-β which have anti-viral actions.
intraepithelial lymphocytes	These are lymphocytes that are integrated into the epithelial layer of the small intestine.
interleukin (IL)	Cytokines produced by leukocytes such as IL-1β, IL-2 and IL-6.
irritable bowel syndrome	Disorder that interferes with the normal function of the colon causing abdominal cramps, bloating, constipation or diarrhoea.
isotype	Refers to the class of immunoglobulin, that is, IgM, IgG, IgA, IgE and IgD.
isotype switching	A molecular process where a class of immunoglobulin is changed to another class. The constant region gene of the heavy chain is changed to a different heavy chain constant region gene by somatic recombination. The antigen specificity is unchanged.
linoleic acid	An essential fatty acid found in vegetables and nut oils. It is an ω-6 fatty acid.
lymph	Fluid-containing proteins and cells that are carried by lymphatic system.
lymph node	Small round structures found at sites where lymphatic vessels converge.
lymphatic vessel	Thin-walled vessels that carry lymph from tissues to the secondary lymphoid tissues (except the spleen) and then to the thoracic duct to join the blood circulation.

(*Continued*)

lymphocyte	A class of white blood cells that consists of B and T lymphocytes of the adaptive immune system and natural killer cells which are large granular cells and are considered lymphocytes of innate immunity.
macrophage	Large mononuclear phagocytic cells that are derived from blood monocytes. They act as antigen presenting cells and phagocytic scavenger cell and are effector cells in adaptive immunity when activated.
metabolic syndrome	A group of risk factors that increases the likelihood of heart disease, Type II diabetes mellitus and stroke. These risk factors are increased blood pressure, high blood glucose, excess fat around the waist, high triglyceride levels and low levels of high density lipoproteins.
microbiota	Microorganisms that reside in the body such as in the gut. They normally do not cause disease and can be beneficial.
MHC	Abbreviation for major histocompatibility complex. MHC Class I and Class II molecules are proteins that are associated with antigenic peptides for presentation to the T cell receptor.
mesenteric lymph node	Lymph nodes in the mesentery (membrane that holds the gut together).
monocyte	White blood cell that is a precursor for tissue macrophages.
monosaccharide	Simple carbohydrate consisting of one sugar molecule. Most common form is glucose.
mucosa	Epithelium that secretes mucus and lines the gastrointestinal, urogenital and respiratory tracts.
naïve T cells	Mature T cells that have left the thymus but not yet encountered specific antigens.
natural killer cell	A lymphocyte that plays an important role in innate immunity. It circulates in blood and has cytotoxic ability against virally-infected cells and certain tumour cells.

(*Continued*)

(Continued)

neutropenia	Abnormally low numbers of neutrophils in blood.
oligosaccharide	Complex carbohydrate that contains 3–10 monosaccharides.
opsonisation	Process of coating the surface of a pathogen with antibody or complement to enhance phagocytosis by neutrophils and macrophages.
pathogen	A microorganism such as bacteria, virus or parasite that causes disease.
Peyer's patch	Gut-associated lymphoid tissues at the wall of the small intestine.
plasma cell	A B lymphocyte that has terminally differentiated and produces antibodies.
prebiotics	Indigestible dietary fibres that serve as food for the 'beneficial'.
probiotics	Live microorganisms that provide beneficial effects by modulating the immune function in the gastrointestinal tract and distant tissues via the mucosal immune system.
polysaccharide	A complex carbohydrate that consists of long chains of glucose.
reactive oxygen species	An oxygen molecule that has become a free radical which is a highly unstable atom with an unpaired electron in its outermost shell.
regulatory T cell	A subset of $CD4^+$ T cell that suppresses or limits the immune response.
retinol	An active alcohol form of Vitamin A that plays an important role in healthy vision and immune function.
retinoic acid	An active acid form of Vitamin A which plays an important role in cell growth and immune function.
SLE	Abbreviation for systemic lupus erythrematosus, an autoimmune disease in which antibodies against DNA, RNA and nucleoproteins form immune complexes to damage blood vessels.

(*Continued*)

starch	Storage form of glucose in plants.
Th1 cell	A subset of CD4 T cells that produces cytokines that promote macrophage activation.
Th2 cell	A subset of CD4 T cells that is characterised by production of non-inflammatory cytokines.
Th17	A subset of CD4 T cells that is characterised by the production of IL-17 and promotion of inflammatory responses.
thymus	A primary lymphoid organ for T cell development. It is located in the upper part of the chest beneath the breast bone.
TNF	Abbreviation for tumour necrosis factor. A cytokine produced by macrophages and T cells and has several actions such as pro-inflammatory effects.
Type I diabetes	An autoimmune disease in which insulin-producing pancreatic β-islet cells are damaged.
Type II diabetes	A long-term chronic condition in which the level of blood glucose is high due to body cells becoming less responsive to insulin.
ulcerative colitis	A chronic disease of the large intestine or colon caused by inflammation or ulceration of the mucosa or innermost lining of the colon.

Index

www.ingramcontent.com/pod-product-compliance
Lightning Source LLC
Chambersburg PA
CBHW060256220326
41598CB00027B/4131